SpringerBriefs in Applied Sciences and Technology

Thermal Engineering and Applied Science

Series Editor

Francis A. Kulacki, Department of Mechanical Engineering, University of Minnesota, Minneapolis, MN, USA

More information about this series at http://www.springer.com/series/10305

Sujoy Kumar Saha • Hrishiraj Ranjan
Madhu Sruthi Emani • Anand Kumar Bharti

Electric Fields, Additives and Simultaneous Heat and Mass Transfer in Heat Transfer Enhancement

 Springer

Sujoy Kumar Saha
Mechanical Engineering Department
Indian Institute of Engineering
Science and Technology, Shibpur
Howrah, West Bengal, India

Hrishiraj Ranjan
Mechanical Engineering Department
Indian Institute of Engineering
Science and Technology, Shibpur
Howrah, West Bengal, India

Madhu Sruthi Emani
Mechanical Engineering Department
Indian Institute of Engineering
Science and Technology, Shibpur
Howrah, West Bengal, India

Anand Kumar Bharti
Mechanical Engineering Department
Indian Institute of Engineering
Science and Technology, Shibpur
Howrah, West Bengal, India

ISSN 2191-530X   ISSN 2191-5318   (electronic)
SpringerBriefs in Applied Sciences and Technology
ISSN 2193-2530   ISSN 2193-2549   (electronic)
SpringerBriefs in Thermal Engineering and Applied Science
ISBN 978-3-030-20772-4   ISBN 978-3-030-20773-1   (eBook)
https://doi.org/10.1007/978-3-030-20773-1

© The Author(s), under exclusive license to Springer Nature Switzerland AG 2020
This work is subject to copyright. All rights are reserved by the Publisher, whether the whole or part of the material is concerned, specifically the rights of translation, reprinting, reuse of illustrations, recitation, broadcasting, reproduction on microfilms or in any other physical way, and transmission or information storage and retrieval, electronic adaptation, computer software, or by similar or dissimilar methodology now known or hereafter developed.
The use of general descriptive names, registered names, trademarks, service marks, etc. in this publication does not imply, even in the absence of a specific statement, that such names are exempt from the relevant protective laws and regulations and therefore free for general use.
The publisher, the authors, and the editors are safe to assume that the advice and information in this book are believed to be true and accurate at the date of publication. Neither the publisher nor the authors or the editors give a warranty, express or implied, with respect to the material contained herein or for any errors or omissions that may have been made. The publisher remains neutral with regard to jurisdictional claims in published maps and institutional affiliations.

This Springer imprint is published by the registered company Springer Nature Switzerland AG
The registered company address is: Gewerbestrasse 11, 6330 Cham, Switzerland

Contents

Nomenclature

d_i	Plain tube inside diameter, or diameter to the base of internal fins or roughness, m or ft
d_o	Tube outside diameter, fin root diameter for a finned tube, m or ft
E	Electric field strength, V/m or V/ft
F_e	Electric field body force, N or lbf
h	Heat transfer coefficient, W/m²-K or Btu/hr-ft²-°F
H	Distance between an electrode and a tube, m or ft
Re_d	Reynolds number based on the tube diameter $= Gd/\mu$, $d = d_i$ for flow inside tube and $d = d_o$ for flow outside tube, dimensionless
Re_{Dh}	Reynolds number based on the hydraulic diameter $= GD_h/\mu$, dimensionless
Re_l	Condensate Reynolds number $(4\Gamma/\mu_l)$, dimensionless
Re_L	Condensate Reynolds number leaving vertical tube $(4\Gamma_L/\mu_l)$, dimensionless
Nu	Nusselt number, dimensionless
L	Tube length, m or ft
r	Radial coordinate, m or ft
W	Condensate flow rate leaving tube, kg/s or lbm/s
u	Fluid velocity, m/s or ft/s
V	Electric potential
a	Ackermann correction factor $(q_{lat}c_p/i_{gv}h_g)$, dimensionless
A	Heat transfer surface area on one side of a direct transfer-type exchanger
A_c	Flow cross-sectional area in minimum flow area, m² or ft²
A_{fr}	Airflow frontal area, m² or ft²
A_o	Total air-side heat transfer surface area, m² or ft²
c_p	Specific heat of fluid at constant pressure, J/kg-K or Btu/lbm-°F
D	Diffusion coefficient, m²/s or ft²/s
D_h	Hydraulic diameter of flow passages, $4LA_c/A$, m, or ft
e	Fin height or roughness height, m or ft
G	Mass velocity based on the minimum flow area, Ga (air), G_w (water), kg/m²-s or lbm/ft²-s

G_{fr} Mass velocity based on flow frontal area, $G_{fr,a}$ (air), $G_{fr,w}$ (water), kg/m^2-s or lbm/ft^2-s

h_g Heat transfer coefficient, W/m^2-K or Btu/hr-ft^2-°F

ITD   Temperature difference between entering fluid temperatures, K or °F

K_p Mass transfer coefficient based on partial pressure driving potential, kg/s-m^2 or lbm/s-ft^2

K_w Mass transfer coefficient based on specific humidity or enthalpy driving potential, kg/s-m^2 or lbm/s-ft^2

L_p Strip flow length of OSF or louver pitch of louver fin, m or ft

M Molecular weight of mixture, kg/kmol or lbm/kmol

N Number of tube rows in the flow direction, or number of tubes in heat exchanger, dimensionless

Sc Schmidt number, v/D, dimensionless

S_l Transverse tube pitch, m or ft

St Stanton number $= h/Gc_p$, dimensionless

T Temperature, Ts (saturated), T_w (wall), T_{vb} (of bulk vapor), T_i (interface), T_{av} (average), T_{el} (entering air), T_{wl} (entering water) K or °F

u Local fluid velocity, m/s or ft/s

W   Specific humidity, dimensionless

W   Fluid mass flow rate $= \rho u_m A c$, kg/s or lbm/s

y   Mass fraction of diffusing component in vapor, dimensionless

y   Coordinate distance normal to wall, m or ft

i Enthalpy of steam-air mixture, kJ/kg or Btu/lbm

i_{gv} Enthalpy of saturated vapor, kJ/kg or Btu/lbm

C_p Particle volume concentration, m^{-3} or ft^{-3}

d_p Particle diameter, m or ft

e Rib height, m or ft

e_o External fin height, m or ft

E_{ho} Heat transfer enhancement ratio ($hA/h_S A_S$ relative to plain surface), dimensionless

G_{mf} Minimum fluidization mass velocity, kg/s-m^2 or lbm/s-ft^2

s Fin spacing, m or ft

x_m Particle-fluid mass fraction ($= x_v \rho_m / \rho_p$), dimensionless

x_v Particle-fluid volume fraction, dimensionless

Greek Symbols

η_f Fin efficiency, dimensionless

μ Fluid dynamic viscosity coefficient, Pa-s or lbm/s-ft

ρ Fluid density, kg/m^3 or lbm/ft^3

α $(1 - e^{-1})/a$, dimensionless

δ Liquid film thickness, m or ft

ν Kinematic viscosity, m^2/s or ft^2/s

β Condensate flooding angle (defined in Figure 12.5), degree
$\varGamma$ Condensate flow rate, per unit tube length, leaving horizontal tube, kg/s-m or lbm/s-ft
ε Fluid permittivity (dielectric constant of fluid), dimensionless
$\varGamma_L$ Condensate flow rate, per unit plate width, leaving vertical plate, kg/s-m or lbm/s-ft
σ_e Electrical conductivity, A/V-m or A/V-ft
ρ_c Electric field space charge density, C/m^3 or C/ft^3

Subscripts

m   Solids-gas or solids-liquid mixture
o   Flow without solids
p   Particles
s   Smooth or plain surface
al   Entering air
b   In bulk vapor
i   At liquid-gas interface
v   Vapor
wl   Entering water
g   Vapor
l   Liquid

Chapter 1
Introduction

Heat transfer can be enhanced by using electric fields, including electrodynamic (EHD) enhancement. A lot of work has been reported on this, Ohadi (1991), Yabe et al. (1987) and Yabe (1991). Yabe (1991) worked on the enhancement of condensation and boiling. Application of an electric field to a dielectric fluid imposes a body force on the fluid. This body force adds a term to the momentum equation and influences the fluid motion. It is important that the fluid be a dielectric and does not conduct any electric current. The body force used in the Navier–Stokes equations is described by the change of Helmholtz free energy for virtual work with the energy stored in the fluid by the electric fluid.

The electric field strength varies with distance from the electrode and depends on the electrode design. Three forces are active here. Electrophoretic (Coulomb) force acts on the net free charge of electric field and space charge density, and the direction of the force depends on the relative polarities of the free charges and the electric field. The dielectrophoretic force is produced by the spatial change of permittivity (also called the dielectric constant). In two-phase flow, this force arises from the difference in permittivity of the vapour and liquid phases. The electrostriction force is caused by inhomogeneity of the electric field strength. This force is dependent on inhomogeneities in the electric field strength and may be likened to an electrical pressure.

Thus, the electric charges are generated by the gradient to the electrical conductivity. Because the electrical conductivity of liquids is temperature dependent, temperature gradients in the fluid generate the electric charge. This makes the Coulomb force an active force.

The momentum and energy equations are coupled through the temperature dependence of permittivity and thermal conductivity of the fluid. Analytical solution of EHD coupled momentum and energy equations is not possible in general, excepting perhaps some simple configurations, Cooper (1992). Rather, data are correlated based on the predicted electric field strength distribution at the heat transfer surface on the basis of uniform electrical fluid properties.

© The Author(s), under exclusive license to Springer Nature Switzerland AG 2020
S. K. Saha et al., *Electric Fields, Additives and Simultaneous Heat and Mass Transfer in Heat Transfer Enhancement*, SpringerBriefs in Applied Sciences and Technology, https://doi.org/10.1007/978-3-030-20773-1_1

Table 1.1 EHD heat transfer enhancement in heat exchanges

References	Maximum enhancement (%)	Test fluid	Wall/electrode configuration	Process
Fremandez and Poulter (1987)	2300	Transformer oil	Tube-wire	Forced convection
Ohadi et al. (1991)	320	Air	Tube-wire/rod	Forced convection
Levy (1964)	140	Silicone oil	Tube-wire	Forced convection
Yabe and Maki (1988)	10,000	R-113 (4% oil)	Plate-ring	Natural convection
Cooper (1990)	1300	R-113 (10% oil)	Tube-wire mesh	Pool boiling
Uemura et al. (1990)	1400	R-113	Plate-wire mesh	Film boiling
Bologa et al. (1987)	2000	Diethyl ether, R-113, hexane	Plate-plate	Film condensation
Sunada et al. (1991)	400:600	R-113; R-123	Vertical wall-plate	Condensation
Ohadi et al. (1992)	480	R-123	Tube-wire	Boiling

The sum of the second and third forces gives the force exerted on dielectric fluids. For the polarization of dielectric fluids, the net charge is zero. However, the force on the polarized charges formed in the stronger field zone is greater than the force on the charges in the weaker field zone. Thus the resultant force (the sum of the forces exerted on each polarized charge) moves fluid elements to the stronger electric field region. Yabe (1991) notes that the effect of the magnetic field generated by the current is negligible compared with the pressure generated by the electric field. Table 1.1 (Ohadi 1991) summarizes enhancement levels obtained by various investigators for natural convection, laminar flow forced convection, boiling and condensation.

Yabe et al. (1978), Velkoff and Godfrey (1979), Yamamoto and Velkoff (1981), Tada et al. (1997), Kasayapanand et al. (2002), Kasayapanand and Kiatsiriroat (2005) and Jiathrakul et al. (2005) all have worked on electrohydrodynamic technique of heat transfer enhancement. The EHD technique was used for two-phase heat transfer by Yabe (1991), Yamashita et al. (1991), Sunada et al. (1991), Bologa et al. (1987), Budov et al. (1987), Grigor'eva and Nakoryakov (1997) and Penev et al. (1968).

Often we encounter with the simultaneous heat and mass transport, most often with two-phase heat transfer involving either condensation or evaporation of mixtures, one of which may be inert at the process conditions. Examples of evaporation processes are vaporization of binary fluids, air humidification (e.g. heat and mass transport in a cooling tower) and desorption of gases from liquid mixtures. Condensation processes include condensation of mixtures, condensation with non-condensable gases, air dehumidification and absorption of vapour in liquid film. Also, the cooling of moisture is an example of simultaneous heat and mass transport. Enhancement techniques are available for this simultaneous heat and mass transport. Again, the impedance for mass transport depends on whether the transport takes place in the gas phase, liquid phase or both. Enhancement techniques may be applied in case of separation of mixtures, which is a classic example of simultaneous heat and mass transport in the realm of chemical engineering. The enhancement of simultaneous transport of heat and mass will be dealt with in this research monograph.

Additives for liquids and gases will be described in detail in this research monograph. Additives for liquids include solid particles or gas bubbles in single-phase flows and liquid trace additives for boiling systems. Additives for gases are liquid droplets or solid particles. This may be either dilute phase (gas-solid suspensions) or dense phase (packed beds and fluidized beds). Wasekar and Manglik (2000), Hetsroni et al. (2001), Tzan and Yang (1990), Wu et al. (1995), Moradian and Saidi (2008), Ammerman and You (1996), Kandlikar and Alves (1999), von Rybinski and Hill (1998) studied the boiling heat transfer enhancement using surfactants.

References

Ammerman CN, You SM (1996) Determination of the boiling enhancement mechanism caused by surfactant addition to water. J Heat Transf 118:429–435

Bologa MK, Savin IK, Didkovsky AB (1987) Electric-field-induced enhancement of vapour condensation heat transfer in the presence of a non-condensable gas. Int J Heat Mass Transf 30:1558–1577

Budov VM, Kir'yanov BV, Shemagin IA (1987) Heat transfer in the laminar-wave section of condensation of a stationary vapour. J Eng Phys 52(6):647–648

Cooper P (1990) EHD enhancement of nucleate boiling. J Heat Transfer 112:458–464

Cooper P (1992) Practical design aspects of EHD heat transfer enhancement in evaporators. ASHRAE Trans 98(Part 2):445–454

Fremandez J, Poulter R (1987) Radial mass flow in electrohydrodynamically-enhanced forced heat transfer in tubes. Int J Heat Mass Transfer 30:2125–2136

Grigor'eva NI, Nakoryakov VE (1997) Exact solution of combined heat mass transfer during film absorption. J Eng Phys 33(5):893–898

Hetsroni G, Zakin JL, Lin Z, Mosyak A, Pancallo EA, Rozenblit R (2001) The Effect of surfactants on bubble growth, wall thermal patterns and heat transfer in Pool boiling. Int J Heat Mass Transf 44:485–497

Jiathrakul W, Kiatsiriroat T, Nuntaphan A (2005) Effect of electric field on heat transfer enhancement in solar air heater. Proc heat- set 2005 Conf Grenoble, France 279–282

Kandlikar SG, Alves L (1999) Effects of surface tension and binary diffusion on Pool boiling of dilute solutions: an experimental assessment. J Heat Transf 121:488–493

Kasayapanand N, Kiatsiriroat T (2005) EHD enhanced heat transfer in Wavy Channel. Int Comm Heat Mass Transf 32:809–821

Kasayapanand N, Tiansuwan J, Asvapoositkul W, Vorayos N, Kiatsiriroat T (2002) Effect of the electrode arrangements in tube Bank on the characteristic of Electrohydrodynamic heat transfer enhancement: low Reynolds number. J Enhanc Heat Transf 9:229–242

Levy E (1964) Effects of electrostatic fields on forced convection heat transfer. M.S. thesis, Massachusetts Institute of Technology, Cambridge, MA

Moradian A, Saidi MS (2008) Electrohydrodynamically enhanced nucleation phenomenon: a theoretical study. J Enhanc Heat Transf 15(1):1–15

Ohadi MM (1991) Heat transfer enhancement in heat exchangers. ASHRAE J 33(12):42–50

Ohadi MM, Nelson DA, Zia S (1991) Heat transfer enhancement of laminar and turbulent pipe flows via corona discharge. Int J Heat Mass Transf 34:1175–1187

Ohadi M, Faani M, Papar R, Radermacher R, Ng T (1992) EHD heat transfer enhancement of shell-side boiling heat transfer coefficients of R-123/oil mixture. ASHRAE Trans 98(Part 2):424–434

Penev V, Krylov SB, Boyadjiev CH, Vorotilin VP (1968) Wavy flow of thin liquid film. Int J Heat Mass Transf 15:1389–1406

Sunada K, Yabe A, Taketani T, Yoshizawa Y (1991) Experimental study of EHD pseudo-dropwise condensation. Proc ASME/JSME Therm Eng 3:61–67

Tada Y, Takimoto A, Hayashi Y (1997) Heat transfer enhancement in a convective field by applying ionic wind. J Enhanc Heat Transf 4:71–86

Tzan YL, Yang YM (1990) Experimental study of surfactant effects on Pool boiling heat transfer. J Heat Transf 112:207–212

Uemura M, Nishiio S, Tanasawa I (1990) Enhancement of pool boiling heat transfer by static electric field. Proc. of the ninth international heat transfer conference. Jerusalem, Israel 4:69–74

Velkoff HR, Godfrey R (1979) Low-velocity heat transfer to a flat plate in the presence of a corona discharge in air. J Heat Trans 101:157–163

von Rybinski W, Hill K (1998) Alkyl Polyglycosides—properties and applications of a new class of surfactants. Angewandte Chemie-International 37:1328–1345

Wasekar VM, Manglik RM (2000) Pool boiling heat transfer in aqueous solutions of an anionic surfactant. Trans ASME J Heat Transf 122:708–715

Wu W-T, Yang Y-M, Maa J-R (1995) Enhancement of nucleate boiling heat transfer and depression of surface tension by surfactant additives. J Heat Trans 117(2):526–529

Yabe A (1991) Active heat transfer enhancement by applying electric fields. In: Proc. third ASME/JSME thermal Eng. C01~f. 3, xv–xxiii

Yabe A, Maki N (1988) Augmentation of convective and boiling heat transfer by applying an electrohydrodynamical liquid jet. Int J Heat Mass Transf 31:407–417

Yabe A, Taketani T, Kikuchi K, Mori Y, Hijikata K (1987) Augmentation of condensation heat transfer around vertical cooled tubes provided with helical wire electrodes by applying nonuniform electric fields. In: Wang B-X (ed) Heat transfer science and technology. Hemisphere, Washington, DC, pp 812–819

Yabe A, Mori Y, Hijikata K (1978) EHD study of the Corona wind between wire and plate electrodes. AIAA J 16:340–345

Yamamoto T, Velkoff HR (1981) Electrohydrodynamics in an electrostatic precipitator. J Fluid Mech 108:1–18

Yamashita K, Kumagai M, Sekita S, Yabe A, Taketani T, Kikuchi K (1991) Heat transfer characteristics on an EHD condenser. Proc ASME/JSME Thermal Eng 3:61–67

Chapter 2
Electrode Design and Its Placement, Enhancement of Single-Phase Gas and Liquid Flow, Theoretical Studies

The design and location of the electrodes depend on whether the flow is single- or two-phase. Also, they depend on the thermophysical and electrical properties of the fluid. The position of the electrodes must be such as that the electric field force aids the hydrodynamic forces, particularly at the heat transfer surface. Separated two-phase flow has distinctly different permittivity of the liquid and vapour. This causes a singularity at the liquid–vapour interface. Jones (1978) has observed that the fluid component having the highest permittivity or dielectric constant moves to the region of the highest field strength. Liquid drop moves towards the electrode, and bubble moves towards the region of lower field strength.

Electrodes are placed at different places depending upon the tube-side and shell-side heat transfer enhancement. If the fluid flows inside a tube, a wire electrode is placed in the centre of the tube for heat transfer as shown in Fig. 2.1. The field strength decays with radial distance from the central electrode wire. If a liquid–gas mixture flows in the tube, the liquid phase will have permittivity. So, the liquid will move towards the central electrode, leaving the vapour at the tube wall. Since the electrode pools condensate away from the condensing surface, condensation enhancement would occur. Yabe et al. (1992) used a perforated fin wrapped in a cylindrical form as the electrode for the enhancement of tube-side convective evaporation of refrigerants. Figures 2.2 and 2.3 show electrode designs used by Cooper (1992) and Yabe et al. (1987), respectively. Cooper (1992) worked for the enhancement of shell-side boiling in a tube bundle. Yabe et al. (1987) worked for the enhancement of condensation on the outer surface of vertical tubes. Figure 2.3 show that electrodes produce a non-uniform electric field. Enhancement occurs irrespective of whether the voltage is AC or DC.

Enhancement increases by increasing the applied voltage. The electrical breakdown strength of the liquid and vapour must be known, and the actual values must not increase the design values. There should be a careful design of the solid insulation system, and poor design may cause a complete electrical breakdown which will cause the EHD enhancement system useless. The insulation material

© The Author(s), under exclusive license to Springer Nature Switzerland AG 2020
S. K. Saha et al., *Electric Fields, Additives and Simultaneous Heat and Mass Transfer in Heat Transfer Enhancement*, SpringerBriefs in Applied Sciences and Technology, https://doi.org/10.1007/978-3-030-20773-1_2

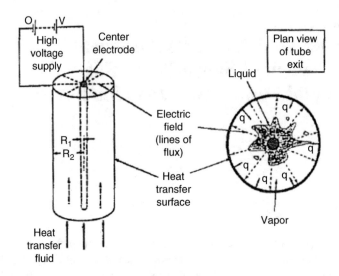

Fig. 2.1 Wire electrode in the centre of tube with two-phase flow from Cooper (1992)

Fig. 2.2 Electrode designs used by Cooper (1992) for the enhancement of shell-side boiling in a tube bundle from Cooper (1992)

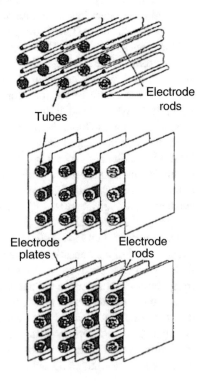

Fig. 2.3 Electrode designs used by Yabe et al. (1987), for the enhancement of condensation on the outer surface of vertical tubes from Yabe et al. (1987)

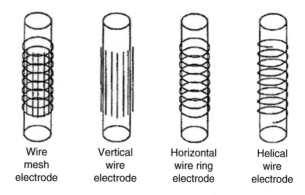

| Wire mesh electrode | Vertical wire electrode | Horizontal wire ring electrode | Helical wire electrode |

depends on the heat transfer media. Polytetrafluoroethylene (PTFE) and epoxy composites are good insulators for this purpose. Electrode and insulation surfaces should be rounded, and conducting rods of large radii should be used to reduce local electric field strength; this was the suggestion made by Cooper (1992) since he noticed that high electric stress exists near small radii of curvature of solids. Kuffel and Zaengl (1984) may be referred for the design aspects of the insulation system on power transformers and switch gear. Small currents minimize physical hazards. All high-voltage surfaces are encased inside the heat exchanger, which is electrically insulated from the shell and tube sides. Adequate insulation has to be used to avoid the possibility of electric shock.

Laohalertdecha et al. (2007) reviewed the electrohydrodynamic techniques required for the enhancement of heat transfer. This active technique requires coupling of an electric field with the fluid field applied in a dielectric fluid medium. In this technique, DC or AC high-voltage low current is typically used. The EHD force (f_e) developed due to flow of charge between charged and grounded electrode is presented as

$$f_e = qE - \frac{1}{2}E^2(\nabla\varepsilon) + \frac{1}{6}\varepsilon_o(K-1)(K+2)(\nabla E)^2$$
$$+ \frac{1}{6}\varepsilon_o E^2 \nabla[(K-1)(K+2)] \tag{2.1}$$

where K is relative permittivity, ε and ε_0 are electric permittivity of fluid and vacuum, respectively. They reviewed the EHD enhancement technique in both single-phase transfer and two-phase transfer. They provided two tables: the first one was for EHD enhancement of condensation (Table 2.1) and the second one refers to EHD enhancement of boiling (Table 2.2). This review presents a better idea for EHD techniques for condensation in horizontal tube and boiling in horizontal and vertical tubes, but no literature has reviewed condensation in vertical tube using EHD techniques.

Dulikravich and Colaco (2004) presented convective heat transfer characteristics under the influence of magnetic and electric fields. For many years, the application of

Table 2.1 EHD enhancement of condensation (Laohalertdecha et al. 2007)

Source	Working fluid	h_{EHD}/h_o (max)
Bologa and DidKovsky	Diethyl ether	20
	R-l13, hexane	10
	Diethyl ether	20
Bologa et al.	Hexane-carbon dioxide	5
Bologa et al.	Freon-113	~20
Bologa et al.	R-l13-helium	1.8
	R-l13-air	2.6
	R-l13-carbon dioxide	3.0
	Hexane-air	5.0
Butrymowicz et al.	R-123	~2
Cheung et al.	R-l34a	7.2
Choi	Freon-R-113	2
Choi and Reynold	R-l13	2
Cooper and Allen	R-12 and R-114	2.9
Damianidis et al.	R-l14	I.08
Didkovsky and Bologa	Non-polar and polar dielectrics	20
Holmes and Chapman	R-l14	Up to 10 times
Jia-Xiang et al.	Freon-11	~1.85
Seth and Lee	R-l13	1.6
Silva et al.	R-l34a	~3
Singh et al.	R-l34a	6.5
Smirnov and Lunev	R-l13, diethyl ether	3.6
Sunada et al.	R-123	~6
Trommelmans and Berghmans	R-l1, R-13 and R-l14	1.1
Velkoff and Miller	Freon-R-113	1.5
Wawzyniak and Seyed-Yagoobi	R-l13	6.1
Yabe et al.	Water, R-l13	2.24
Yabe et al.	Silicone oil, R-l13	4.5
Yamashita et al.	$C_6 F_{14}$ (perfluorohexane)	6
	R-l14	6
	n-Perfluorohexane	4

electric and magnetic fields has been used in magnetohydrodynamic pumping, electrohydrodynamic pumping and electromagnetic stirring of molten metal. The parameters for Navier–Stokes equations and Maxwell's equations for a standard MHD model have been tabulated in Table 2.3. Similarly, the corresponding parameters for EHD models have been listed in Table 2.4. The various optimization models of magnetic and electric fields available are differential evolution algorithm, quasi-Newtonian Pshenichny-Danilin algorithm, genetic algorithm, Davidon–Fletcher–Powell gradient search algorithm and modified Nelder–Mead simplex algorithm.

Table 2.2 List of the EHD enhancement of boiling (Laohalertdecha et al. 2007)

Source	Working fluid	h_{EHD}/h_0(max)
Allen and Cooper	R-114	Up to 60
Blachowicz et al.	Benzene	2
Bochirol et al.	Trichlorethylene and ethyl ether	2.7
	Methyl alcohol	6
Cheung et al.	R-134a	5.1
Cooper	R-114 + oil	13
Damianidis et al	R-114 (smooth tube)	~1
	R-114 (low-fin tubes)	2.5
Feng and Seyed-Yagoobi	R-134a	~4.5
Kawahira et al.	R-11	4
Karayiannis	R-123 and R-11	9.3
Karayiannis et al.	R-114	4
Liu et al.	R-123	2.1
Neve and Yan	R-114	3
Ohadi et al.	R-123	5.5
	R-11	1.7
Ogata et al.	R-123	7
Papar et al.	R-123	6
Schnurmann and Lardge	n-Heplane	~7
	20% solution of isopropyl alcohol in n-heptane	~7
	Perfluoromethylcyclohexane	~2
Watson	n-Hexane	2.6
Source	R-123 + R-134a	3
Yabe et al	R-123	6
Yamashita and Yabe	Acetone, benzene	1.82
Zheltukhin et al.	n-Diethyl ether	(acetone)

Table 2.3 Parameters for the Navier–Stokes and Maxwell's equations in a standard MHD model (Dulikravich and Colaco 2004)

Conservation of	λ	ϕ	ϕ^*	ϕ^{**}	ϕ^{***}	Γ	S
Mass	ρ	1	1	1	1	**0**	**0**
x-Momentum	ρ	u	u	u	u	μ	$-\dfrac{\partial p}{\partial x} - \dfrac{B_y}{\mu_m}\left[\dfrac{\partial B_y}{\partial x} - \dfrac{\partial B_x}{\partial y}\right]$
y-Momentum	ρ	v	v	v	v	u	$-\dfrac{\partial p}{\partial y} - \rho g \left[1 - \beta(T - T_o)\right]$ $+\dfrac{B_x}{\mu_m}\left[\dfrac{\partial B_y}{\partial x} - \dfrac{\partial B_x}{\partial y}\right]$
Energy	ρ	h	h	h	T	k	$\dfrac{1}{\sigma\mu_m^2}\left[\dfrac{\partial B_y}{\partial x} - \dfrac{\partial B_x}{\partial y}\right]^2$
Magnetic flux in x-direction	1	B_x,	**0**	B_x	B_x	$\dfrac{1}{\mu_m\sigma}$	$\dfrac{\partial(uB_y)}{\partial y}$
Magnetic flux in y-direction	1	B_y	B_y	**0**	B_y	$\dfrac{1}{\mu_m\sigma}$	$\dfrac{\partial(uB_x)}{\partial x}$

Table 2.4 Parameters for the Navier–Stokes and Maxwell's equations in a standard EHD model (Dulikravich and Colaco 2004)

Conservation of	λ	ζ	ϕ	ϕ^*	ϕ^{**}	ϕ^{***}	Γ	S
Mass	ρ	0	1	1	1	1	0	0
x-Momentum	ρ	0	u	u	u	u	μ	$-\dfrac{\partial p}{\partial x} + q_e E_x$
y-Momentum	ρ	0	v	v	v	v	μ	$-\dfrac{\partial p}{\partial y} - \rho g\left[1 - \beta(T - T_o)\right] + q_e E_y$
Energy	ρ	0	h	h	h	T	k	$q_e\left[b\left(E_x^2 + E_y^2\right) + uE_x + vE_y\right]$ $-D_e\left(E_x\dfrac{\partial q_e}{\partial x} + E_y\dfrac{\partial q_e}{\partial y}\right)$
Electric potential	0	0	0	0	0	φ	-1	$\dfrac{q_e}{\varepsilon_0}$
Electric charged particle transport	1	b	q_e	q_e	q_e	q_e	D_e	0

Dulikravich et al. (1993) and Dennis and Dulikravich (2000) presented similar works. Lee et al. (1991), Sabhapathy and Salcudean (1990), Fedoseyev et al. (2001), Motakeff (1990), Eringen and Maugin (1990), Ko and Dulikravich (2000), Sampath and Zabaras (2001), Dennis and Dulkravich (2001, 2002) and Colaço et al. (2003, 2004a, b) presented more information on heat transfer augmentation using electric and magnetic fields.

Trommelmans et al. (1985) investigated experimentally as well as theoretically the effect of uniform electric field applied perpendicularly to the plate for the flow of R-11, R-113 and R-114 refrigerant fluids. They proposed correlation for the heat transfer coefficient. Similarly, Bologa et al. (1996) investigated and proposed correlations for heat transfer coefficient considering the effect of electric field. Jia-Xiang et al. (1996) used double-electrode cylinder heat transfer model for Freon-11 and studied EHD-coupled heat transfer system. They found that electric field showed positive effect on heat transfer for condensation as well as boiling. Cao et al. (2003) investigated the drying characteristics of wheat with the application of high-voltage electrostatic field (HVEF) and found that the rate of drying increased with increased voltage and decreased discharge gap accordingly with the negligible power consumption.

Bologa and Didkovesky (1977) experimented on film condensation with a non-uniform electric field whose strength and frequency varied. They used n-hexane, R-113 and diethyl ether for film condensation heat transfer on vertical flat plate and tube and proposed some explicit correlation. Later, Bologa et al. (1987a, b, 1995) worked on vapour–gas mixture and two-phase system for various shapes of interface including the role of capillary process on a vertical plate for film condensation heat transfer enhancement. Smirnov and Lunev studied non-polar (R-113) and weakly polar fluid (diethyl ether) for predicting the heat transfer coefficient with DC and AC electric fields on a vertical tube. Didkovsky and Bologa (1981) experimented on film condensation of a stagnant pure vapour on vertical surface and concluded that 20 times increase in heat transfer coefficient was achieved. Also, Didkovsky and Bologa (1981) correlated the heat transfer coefficient with film thickness ratio for outside vertical plates which were under gravitational and EHD forces.

Grassi et al. (2007) studied the application of EHD technique for heat transfer augmentation in single-phase flow. This results in the generation of ions which move towards the electrode with opposite charge. In this process, the ions drag the fluid molecules that are neutral in nature. When this electrohydrodynamic flow comes in contact with the heating surface, there is scope for large amount of heat transfer. In order to achieve augmented heat transfer rate at less pumping power, optimization is required. More information on EHD augmentation technique can be obtained from Grassi et al. (2007). The EHD parameters that affect the enhancement are geometrical parameters of the electrode such as shape, orientation, radius of curvature and micro-asperities present in the flow such as polarity of the electrode and properties of the fluid.

The Nusselt numbers for liquids, air and some metals are shown in Figs. 2.4 and 2.5. The Nusselt numbers for natural convection and EHD technique have been compared. They observed that the Nusselt number for liquids mainly depend on the composition and polarity of the electrode and the fluid properties. The probability of oxidation at anode or reduction at cathode depends on the metal work function, active solvated impurity type and their correlation. The ionization process actually involves neutral extrinsic molecules; hence, the performance of polar liquids like HFE 7100 and Vertrel®XF is better than that of non-polar liquids, FC-72. The

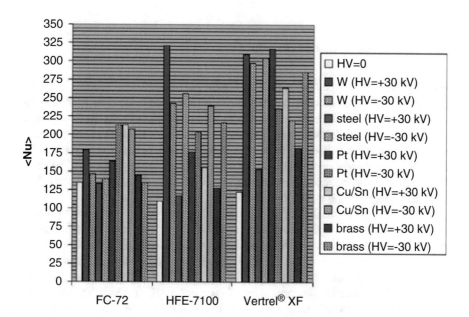

Fig. 2.4 Nusselt number for three liquids and five metals (Grassi et al. 2007)

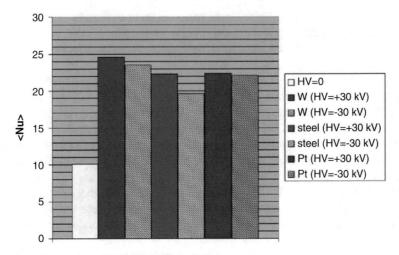

Fig. 2.5 Nusselt number for air and three metals (Grassi et al. 2007)

enhancement techniques applied for natural convection of air flow also yielded quite low enhancement rates. This is attributed to the fact that the mean free path of air is very long due to which the momentum transfer is reduced globally resulting in lower enhancement. Also, for metals the intense positive corona winds rather than negative winds were observed. They concluded that HFE 7100 usage with negative voltage applied to stainless steel sharp electrode resulted in optimal performance.

2.1   Enhancement of Gas Flow

Electrophoretic force may cause EHD-induced fluid motion known as ionic (or corona) wind effect for single-phase flows. Ions are produced close to the surface of a wire anode. The Coulomb force on the ions is responsible for the ions to move to the cathode surface. Interaction of the corona wind with the main flow produces mixing by secondary flows, and greatest enhancement occurs for low Reynolds number laminar natural convection. With the increase in the main flow velocity, turbulent flow occurs, and the secondary flow speed is insignificant in comparison with the strength of the turbulent eddies.

Figure 2.6 shows the forced convection enhancement data observed by Ohadi et al. (1991a). They used single- and double-axial electrodes to measure the enhancement of air flow. Ohadi et al. (1991b) used axial wire electrodes in a double-pipe heat exchanger with air flow in both streams. EHD enhancement data were given by them for tube-side, shell-side and both fluids. Ohadi et al. (1994) extended the work to a shell and tube heat exchanger consisting of seven tubes. $1000 < Re < 6000$ was used for the tube and the shell sides. More than 100% enhancement was obtained at the lowest Reynolds numbers when both sides were excited. With the increase of

Fig. 2.6 Enhancement
provided by axial single-
and double-wire electrodes
in a 38.4-mm-diameter tube
from Ohadi et al. (1991b)

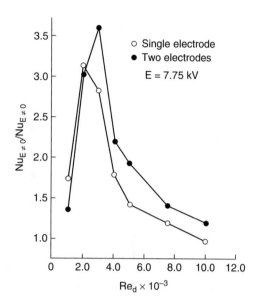

Reynolds number, however, the enhancement decreased. Nelson et al. (1991) studied EHD enhancement of air flow in circular tube using a central, axial wire electrode. They observed almost similar increase in pressure drop and heat transfer. Blanford et al. (1995) applied EHD technique to enhance air-side heat transfer of a one row finned tube heat exchanger. Munakata et al. (1993) and Molki et al. (2000) observed reduced frost formation by EHD under natural convection environment.

Kasayapanand et al. (2006) studied electrodynamic technique for heat transfer in a solar air heater with double-flow configuration. Laminar forced heat convection was considered for the study. A solar air heater with double-flow configuration has been shown in Fig. 2.7. Five different types of arrangements of electrodes have been presented in Fig. 2.8. For each configuration, 20 electrodes have been considered. Ten electrodes on each of upper and lower channels (Case 1), 20 electrodes on upper channel (Case 2), 20 electrodes on lower channel (Case 3) and 15 and 5 electrodes on upper and lower channels (Cases 4 and 5) are the different configurations of electrode arrangements used in the study. A uniform distribution of space between electrodes is maintained along the tube length. They observed rotating cellular motion around all the electrodes in Cases 1–3. But, as the number of electrodes is high in Cases 2 and 3 as compared to that in Case 1, more vortices around the electrode were observed. Also, oscillating flow patterns were reported because of high-density electric field.

The number of vortices along the tube length was found to be varying in Cases 4 and 5, due to varying density of electrodes along the tube length. The efficiency of the solar collector with double flow has been shown in Fig. 2.9 for different electrode arrangements. Although high heat transfer coefficients are observed in Cases 2 and 3, with further increase in the number of electrodes, recirculation in the middle zone of the heating plate occurs and decreases the heat transfer coefficients. They

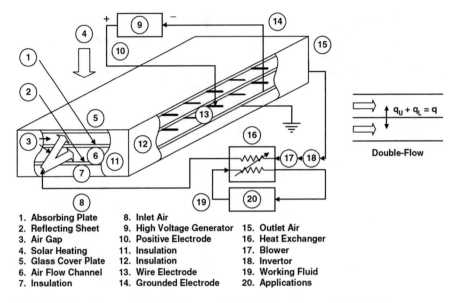

1. Absorbing Plate   8. Inlet Air
2. Reflecting Sheet   9. High Voltage Generator   15. Outlet Air
3. Air Gap   10. Positive Electrode   16. Heat Exchanger
4. Solar Heating   11. Insulation   17. Blower
5. Glass Cover Plate   12. Insulation   18. Invertor
6. Air Flow Channel   13. Wire Electrode   19. Working Fluid
7. Insulation   14. Grounded Electrode   20. Applications

Fig. 2.7 Solar air heater with double-flow configuration (Kasayapanand et al. 2006)

Fig. 2.8 Five different types of arrangements of electrodes (Kasayapanand et al. 2006)

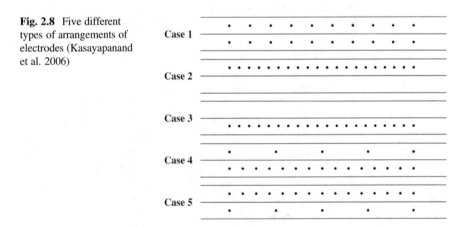

concluded that heat transfer augmentation with electrohydrodynamic technique was observed with increase in supply voltage and decrease in Reynolds number. The configuration with electrodes arranged on the low channel showed the best performance. Also, the lower air channel height showed better heat transfer performance. They proposed the ratio of distance between wire electrodes and distance of channel height to be 1, to achieve improved heat transfer enhancement rates. Satcunanathan and Deonarine (1973), Wijeysundera et al. (1982) and Yeh et al. (1999, 2002) have also worked with double-flow or two-pass arrangements.

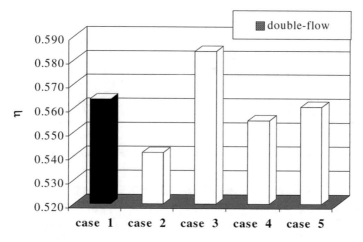

Fig. 2.9 Efficiency of the solar collector with double flow (Kasayapanand et al. 2006)

2.2 Enhancement of Liquid Flow

Fernandez and Poulter (1987) worked with transformer oil flowing on the annulus side of a double-pipe heat exchanger. Ishiguro et al. (1991) worked with a 20 mm high rectangular channel and used axial wire electrodes, and the working fluid was a mix of R-113 and ethanol. Very high enhancement of heat transfer was observed.

Wang and Bao (2011) investigated the effect of electrohydrodynamics on augmentation of heat transfer of transformer oil. They reported that for low values of applied voltage (<4 kV), increase in heat transfer augmentation ratio with increase in applied voltage was observed. This is followed by a slow increase in enhancement ratio which tends to reach a constant value after 4 kV of applied voltage. Also, the increase in friction factor was only nominal for high applied voltage. Thus, the overall improvement in heat transfer and pressure drop characteristics can be achieved by using high-voltage electric field.

Mirzaei and Saffar-Avval (2018) used EHD conduction method for the study of laminar convective heat transfer in an annulus. Due to the wide applications of EHD method, many researchers Bergles (1973), Seyed-Yagoobi (2005), Alamgholilou and Esmaeilzadeh (2012), Laohalertdecha et al. (2007), Liu et al. (2005), Cotton et al. (2012), Laohalertdecha and Wongwises (2006), Sadek et al. (2010), Jalaal et al. (2013), Van Poppel et al. (2010), Kasayapanand and Kiatsiriroat (2009), Grassi et al. (2005a, b), and Lakeh and Molki (2012) worked on pumping, single-phase heat transfer, condensation, mixing and liquid jets with this method. Atten and Seyed-Yagoobi (2003) developed a theoretical model for conduction pumping, and it was propagated by Jeong and Seyed-Yagoobi (2004), Feng and Seyed-Yagoobi (2004) and Pearson and Seyed-Yagoobi (2009). Mirzaei and Saffar-Avval (2018) experimented with novel double-pipe heat exchanger with the outer pipe having Teflon parts and copper inner pipe.

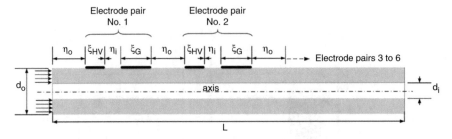

Fig. 2.10 Sketch of the geometry and dimensional parameters (Mirzaei and Saffar-Avval 2018)

Table 2.5 Dimensions of the geometry (Mirzaei and Saffar-Avval 2018)

d_t (mm)	d_o (mm)	L (mm)	ξ_{HV} (mm)	ξ_G (mm)	η_I (mm)	η_o (mm)
9.5	29	494/506/513	6	15	3/5/7	40

Table 2.6 Electrical and thermophysical properties of the working fluid (Mirzaei and Saffar-Avval 2018)

Property	Value [26]
Permittivity (F/m)	18.41×10^{-12}
Electrical conductivity (S/m)	1.2×10^{-12}
Density (kg/m^3)	895.5
Dynamic viscosity (Pa·s)	0.011
Specific heat (J/kg·K)	1856 [35]
Thermal conductivity (W/m·K)	0.13 [35]
Thermal charge mobility (m^2/V·s)	0.0007
Positive charge mobility (m^2/V·s)	0.18×10^{-8}
Negative charge mobility (m^2/V·s)	0.35×10^{-8}

Their objective was to configure the electrodes for heat transfer enhancement in annuli and its mounting. The electrode configurations have been presented in Fig. 2.10 with their dimensions in Table 2.5. They made the following assumptions: (1) for symmetric electrodes, the net flow generation of EHD conduction pump depends on charge mobility, and if the mobility of negative charge is higher than that of positive charge, it would be from cathode to anode; and (2) the net flow generation for identical negative and positive charge mobility would be from narrower electrodes to wider electrodes. They used transformer oil as the working fluid and presented its properties in Table 2.6. The original experimental setup with EHD pumps and testing section including double pipes are shown in Fig. 2.11. The testing section has six EHD pumps, and each pump consists of a PTFE electrode holder. The fluid motion and electric field are governed by Navier–Stokes equations and Maxwell's equations, respectively. They considered steady state incompressible fluid with no charge injection from electrodes. They used finite volume method and PISO algorithm for equation solving and pressure velocity coupling, respectively.

The experimental study and its numerical simulated results were compared, and they found 10% average error from Nusselt number versus voltage plot. They

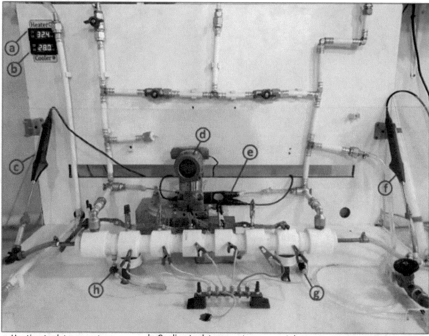

a. Heating tank temperature   b. Cooling tank temperature   c. Inlet water temperature sensor
d. Differential pressure transmitter   e. Oil temperature sensors   f. Outlet water temperature sensor
g & h. Pressure sensor tap

Fig. 2.11 The double pipe with EHD pumps and measurement instruments in the test section with their positions (Mirzaei and Saffar-Avval 2018)

concluded that enhancement in heat transfer was observed with increased Reynolds number. EHD heat transfer enhancement increases by 50% and 21% at Reynolds number 10 and 30, respectively. This shows that the enhancement is effective at lower Reynolds number. They concluded that pressure drop decreased with increase in Reynolds number as well as it was less for 8 kV than that for 16 kV, which is presented in Fig. 2.12. They plotted Nusselt number variation with pumping power (Fig. 2.13a) and the effectiveness parameter versus Reynolds number (Fig. 2.13b). They observed that electric field enhances heat transfer. Net charge density contours near electrode pair number 3 are shown in Fig. 2.14. The increase in voltage and electric force caused formation of two vortices which became prominent with increased voltage. This is the reason behind heat transfer enhancement. Distortion of isotherms at different applied voltages is presented in Fig. 2.15. Thinner boundary layer was observed at higher Reynolds number.

For single-phase flow, Fernandez and Poulter (1987) investigated heat transfer for flow through annular cross-section and presented the experimental results based on EHD technique used for heat transfer enhancement. They claimed that for 30 kV DC voltage across annular gap, the rate of heat transfer was enhanced by 20 times and the pressure drop increased only three times under laminar flow conditions.

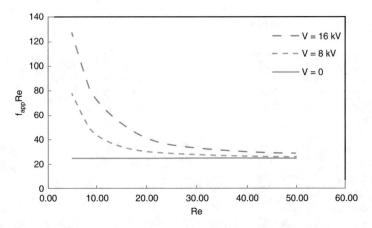

Fig. 2.12 Apparent friction factor as a function of Reynolds number with and without EHD (Mirzaei and Saffar-Avval 2018)

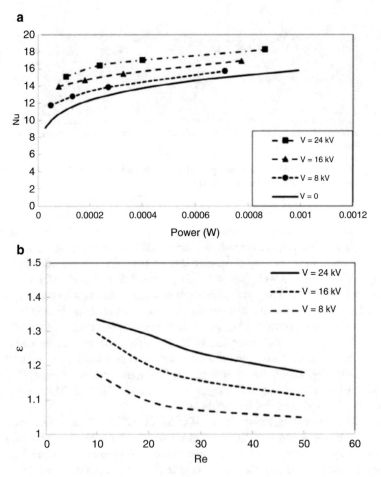

Fig. 2.13 (**a**) Nusselt number as a function of pumping power and (**b**) effectiveness parameter as a function of Reynolds number (Mirzaei and Saffar-Avval 2018)

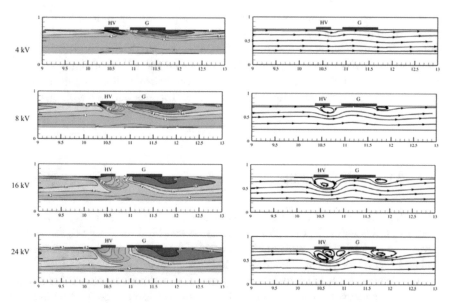

Fig. 2.14 Net charge distribution (left) and flow streamlines (right) in different applied voltages at $Re = 10$ (Mirzaei and Saffar-Avval 2018)

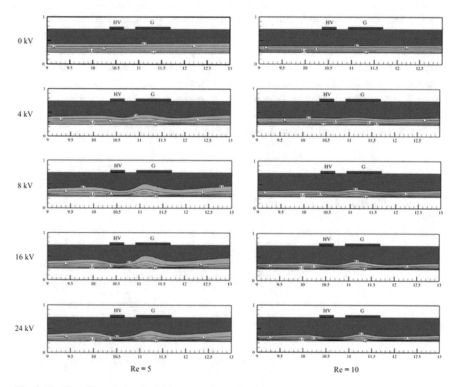

Fig. 2.15 The effect of electric field on non-dimensional isotherms at $Re = 5$ and $Re = 10$ (Mirzaei and Saffar-Avval 2018)

Paschkewitz and Pratt (2000) investigated the pressure drop, heat transfer rate, electrical power requirement and transition between viscous and electrically influenced flow regions by EHD enhancement technique. They used three cooling oils having different fluid properties and examined the effect of fluid properties on EHD enhancement. They observed that lower viscosity liquids suit the EHD technique at relatively low Reynolds number, and it was verified with analytical results.

2.3   Theoretical Studies

Numerical studies on forced convection enhancement in a channel by EHD were done by Hasegawa et al. (1999) for turbulent flow and Molki et al. (2000) for laminar flow. The details of this study may be obtained from these references. A corona-induced secondary flow in the cross-section of the channel was observed.

Terekhov et al. (2018) studied the conjugate heat and mass transfer enhancement for laminar air flow through a channel. They considered forced convection in a cell of direct evaporating cooler under the uniform heat flux. They numerically investigated the effect of Reynolds number ($Re = 50$–1000), temperature ($T_O = 10$–$40\,°C$), humidity of air ($\phi_O = 0$–50%) at the inlet and heat flux range [$q^* = (-0.1)$–$(+0.2)$] on the rate of evaporation. They solved the two-dimensional stationary Navier–Stokes, energy and diffusion equations for laminar regime. They found that by increasing the additional thermal flux to the wall, vapour concentration and temperature of air at the channel exit increased. They pointed that increase of normal vapour mass flux on the surfaces resulted in more effective process of wall drying. They studied the analogy between heat and mass transfer and friction on evaporating surface. Figure 2.16 shows the variation of bulk temperature along the channel for different additional heat flux at a given Reynolds number of 200 and 30 °C inlet

Fig. 2.16 The effect of the additional heat flux on bulk temperature (Terekhov et al. 2018)

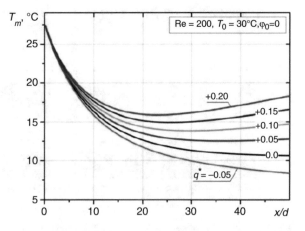

Fig. 2.17 An air cooling parameter vs. additional heat flux and air humidity (Terekhov et al. 2018)

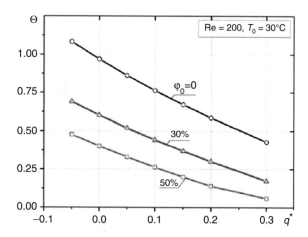

Fig. 2.18 The cooling efficiency of the evaporating cell (Terekhov et al. 2018)

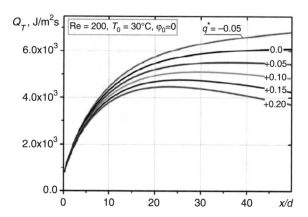

temperature of air. Degree of cooling parameter (Θ) of gas in the evaporator was given by the relation:

$$\Theta = \frac{T_m - T_0}{T_S - T_0} \tag{2.2}$$

Figure 2.17 shows the variation of air-cooling parameter with the addition of heat flux for different values of air relative humidity. It was observed that air-cooling parameter decreased remarkably with increase in relative humidity at the inlet. The cooling efficiency of the evaporating cell was first increasing then decreasing along the channel with decreasing of additional heat flux as shown in the Fig. 2.18. The variation of thermohydraulic efficiency along the channel with different additional heat flux and humidity of air at the inlet of channel is shown in the Fig. 2.19. The numerical results showed that the thermohydraulic efficiency decreased with increasing additional heat flux and relative humidity. Anisimov and Pandelidis (2014), Duan et al. (2012), Debbissi et al. (2008), Nasr et al. (2009) and Terekhov

Fig. 2.19 The parameter of thermohydraulic efficiency vs. additional heat flux (Terekhov et al. 2018)

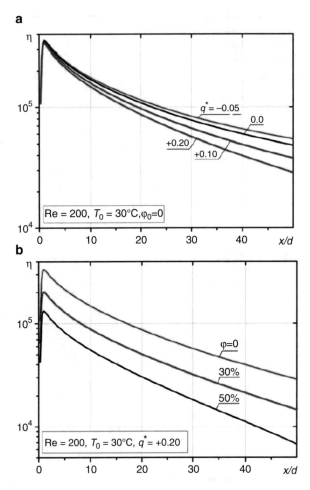

et al. (2015) studied the heat and mass transfer in the evaporative cooling during phase transitions.

Moatimid (1994) examined two fluids having cylindrical interface between them. He targeted to know the effect of periodic tangential force regulated on the cylindrical interface which can transfer both heat and mass. He discussed and derived a simple general dispersion equation which was numerically validated.

Kasayapanand and Kiatsiriroat (2005) used numerical investigation method for the calculation of rate of heat transfer inside wavy channel with different wire electrode arrangements in the laminar flow regime. They used high-voltage DC current for electric field and chose air as working fluid. They considered the electric field, flow filed and temperature field in their mathematical formulation and concluded that heat transfer coefficient increased with decreasing Reynolds number under the influence of electric field. Also, they observed that the wave aspect ratio, number of wave per unit length and number of wire electrodes altered the heat transfer rate.

Ghazi et al. (2011) numerically investigated coupled natural and electroconvective heat transfer characteristics. The computational fluid dynamics have been used to numerically study the effect of electric field, flow field and temperature field interactions. They observed that the number of EHD wire electrodes always did not result in increased heat transfer. The applied voltage has significant effect on the flow field and temperature field. They reported that the enhancement rates were more for low Rayleigh numbers. They concluded that the best heat transfer performance was observed when the number of EHD wire electrodes was equal to the number of Benard cells.

References

Alamgholilou A, Esmaeilzadeh E (2012) Experimental investigation on hydrodynamics and heat transfer of fluid flow into channel for cooling of rectangular ribs by passive and EHD active enhancement methods. Exp Thermal Fluid Sci 38:61–73

Anisimov S, Pandelidis D (2014) Numerical study of the Maisotsenko cycle heat and mass exchanger. Int J Heat Mass Transf 75:75–96

Atten P, Seyed-Yagoobi J (2003) Electrohydrodynamically induced dielectric liquid flow through pure conduction in point/plane geometry. IEEE Trans Dielectr Electr Insul 10:27–36

Bergles AE (1973) Techniques to augment heat transfer. Handbook of heat transfer. A 74–17085 05-33. McGraw-Hill, New York, pp. 10–11

Blanford MD, Ohadi MM, Dessiatoun SV (1995) Compound air-side heat transfer enhancement in a cross-flow refrigerant-to-air heat exchanger. ASHRAE Trans 101(Part 2):1049–1054

Bologa MK, Didkovesky AB (1977) Enhancement of heat transfer in film condensation of vapour of dielectric fluid by superposition of electric fields. Heat Transfer Soviet Res 9:147–151

Bologa MK, Savin IK, Didkovsky AB (1987a) Electric-field-induced enhancement of vapour condensation heat transfer in the presence of a non-condensable gas. Int J Heat Mass Transf 30:1558–1577

Bologa MK, Savin IK, Didkovsky AB (1987b) Electric-field-induced enhancement of vapour condensation heat transfer in the presence of a non-condensable gas. Int J Heat Mass Transf 30:1577–1585

Bologa MK, Korovkin VP, Savin IK (1995) Mechanism of condensation heat transfer enhancement in an electric field and the role of capillary processes. Int J Heat Mass Transf 38:175–182

Bologa MK, Sajin TM, Kozhukhar LA, Klimov SM, Motorin OV (1996) The influence of electric fields on basic processes connected with physical phenomena in two-phase systems. International conference on conduction and breakdown in dielectric liquid 69–72

Cao W, Nishiyama Y, Koide S (2003) Electrohydrodynamic drying characteristics of wheat using high voltage electrostatic field. J Food Eng 62(3):209–213

Colaço MJ, Dulikravich GS, Martin TJ (2003) Reducing convection effects in solidification by applying magnetic fields having optimized intensity distribution. ASME paper HT2003-47308 Las Vegas NV

Colaço MJ, Dennis BH, Dulikravich GS, Martin TJ, Lee SS (2004a) Optimization of intensities, and orientations of magnets controlling melt flow during solidification. Materials Manufacturing Processes 19(4):695–718

Colaço MJ, Dulikravich GS, Martin TJ (2004b) Optimization of wall electrodes for electro-hydrodynamic control of natural convection effects during solidification. Mater Manufact Proc 19(4):719–736

Cooper P (1992) Practical design aspects of EHD heat transfer enhancement in evaporators. ASHRAE Trans 98(Part 2):445–454

Cotton JS, Robinson AJ, Shoukri M, Chang JS (2012) AC voltage induced electro-hydro-dynamic two-phase convective boiling heat transfer in horizontal annular channels. Exp Thermal Fluid Sci 41:31–42

Debbissi C, Orfi J, Nassrallah S (2008) Numerical analysis of the evaporation of water by forced convection into humid air in partially wetted vertical plates. J Eng Appl Sci 3(11):811–821

Dennis BH, Dulikravich GS (2000) Simulation of Magnetohydrodynamics with conjugate heat transfer. In Onate E, Bugeda G, Suarez B (eds) European congress on computational methods in applied sciences and engineering, Barcelona, Spain

Dennis BH, Dulikravich GS (2001) Optimization of magneto-hydrodynamic control of diffuser flows using micro-genetic algorithm and least squares finite elements. J Finite Elements Anal Desn 37(5):349–363

Dennis BH, Dulikravich GS (2002) Magnetic field suppression of melt flow in crystal growth. Int J Heat Fluid Flow 23(3):269–277

Didkovsky AB, Bologa MK (1981) Vapour film condensation heat transfer and hydrodynamics under the influence of an electric field. Int J Heat Mass Transf 24:811–819

Duan Z, Zhan C, Zhang X, Mustafa M, Zhao X, Alimohammadisagvand B, Hasan A (2012) Indirect evaporative cooling: past, present and future potentials. Renew Sust Energ Rev 16:6823–6850

Dulikravich BD, Ahuja V, Lee S (1993) Simulation of electrohydrodynamic enhancement of laminar flow heat transfer. In: Bayazitoglu Y, Arpaci VS (eds) Fundamentals of heat transfer in electromagnetic, electrostatic and acoustic fields., HTD-Vol, vol 248, pp 43–52

Dulikravich GS, Colaco MJ (2004) Convective heat transfer control using magnetic and electric fields. J Enhanc Heat Transf 13(2):139–155

Eringen AC, Maugin GA (1990) Electrodynamics of continua II—fluids and complex media. Springer, New York

Fedoseyev KI, Kansa EJ, Marin C, Ostrogorsky AG (2001) Magnetic field suppression of semi-conductor melt flow in crystal growth: comparison of three methods for numerical modeling. Jpn CFD J 9:325–333

Feng Y, Seyed-Yagoobi J (2004) Understanding of electrohydrodynamic conduction pumping phenomenon. Phys Fluids 16(7):2432–2441

Fernandez J, Poulter R (1987) Radial mass flow in electrohydrodynamically-enhanced forced heat transfer in tubes. Int J Heat and Mass Transf 30:2125–2136

Ghazi R, Saidi MS, Saidi MH (2011) Numerical study of enhanced heat transfer by coupling natural and electro-convections in a horizontal enclosure. J Enhanc Heat Transf 18(6):503–511

Grassi W, Testi D, Saputelli M (2005a) EHD enhanced heat transfer in a vertical annulus. Int Commun Heat Mass Transf 32(6):748–757

Grassi W, Testi D, Saputelli M (2005b) Heat transfer enhancement in a vertical annulus by electrophoretic forces acting on a dielectric liquid. Int J Therm Sci 44(11):1072–1077

Grassi W, Testi D, Della Vista D (2007) Optimal working fluid and electrode configuration for EHD-enhanced single-phase heat transfer. J Enhanc Heat Transf 14(2):161–173

Hasegawa M, Yabe A, Nariai H (1999) Turbulent generation and mechanism analysis of forced convection heat transfer enhancement by applying electric fields in the restricted region near the wall. In: Proc. 5th ASMEIJSME thermal Eng. Joint Conj. paper ATJE99-6380

Ishiguro H, Nagata S, Yabe A, Nariai (1991) Augmentation of forced-convection heat transfer by applying electric fields to disturb flow near a wall. In B-X Wang, (ed) Heat transfer science and technology. Hemisphere, New York, 25–31

Jalaal M, Khorshidi B, Esmaeilzadeh E (2013) Electro-hydro-dynamic (EHD) mixing of two miscible dielectric liquids. Chem Eng J 219:118–123

Jeong SI, Seyed-Yagoobi J (2004) Fluid circulation in an enclosure generated by electrohydrodynamic conduction phenomenon. IEEE Trans Dielectr Electr Insul 11(5):899

Jia-Xiang Y, Li-Jian D, Yang H (1996) An experimental study of EHD coupled heat transfer. IEEE 1:348–351

Jones TB (1978) Electohydrodynamically enhanced heat transfer in liquids—a review. Adv Enhanced Heat Transfer 14:107–148

Kasayapanand N, Kiatsiriroat T (2005) EHD enhanced heat transfer in wavy channel 809–821

Kasayapanand N, Kiatsiriroat T (2009) Enhanced heat transfer in partially open square cavities with thin fin by using electric field. Energy Conver Manag 50(2):287–296

Kasayapanand N, Kiatsiriroat T, Vorayos N (2006) Enhanced heat transfer in a solar air heater with double-flow configuration by electrohydrodynamic technique. J Enhanc Heat Transf 13(1):39

Ko HJ, Dulikravich GS (2000) A fully non- linear model of electro-magneto-hydrodynamics. Int J Non-Linear Mech 35(4):709–719

Kuffel E, Zaengl WS (1984) High-voltage engineering. Pergamon, Oxford

Lakeh RB, Molki M (2012) Targeted heat transfer augmentation in circular tubes using a corona jet. J Electrost 70(1):31–42

Laohalertdecha S, Wongwises S (2006) Effects of EHD on heat transfer enhancement and pressure drop during two-phase condensation of pure R-134a at high mass flux in a horizontal micro-fin tube. Exp Thermal Fluid Sci 30(7):675–686

Laohalertdecha S, Naphon P, Wongwises S (2007) A review of electrohydrodynamic enhancement of heat transfer. Renew Sustain Energ Rev 11(5):858–876

Lee SS, Dulikravich GS, Kosovic B (1991) Electrohydrodynamic (EHD) Flow Modeling and Computations, AIAA Paper 91–1469, AIAA Fluid, Plasma Dynamics and Lasers Conference, Honolulu, Hawaii

Liu Y, Li R, Wang F, Yu H (2005) The effect of electrode polarity on EHD enhancement of boiling heat transfer in a vertical tube. Exp Thermal Fluid Sci 29(5):601–608

Mirzaei M, Saffar-Avval M (2018) Enhancement of convection heat transfer using EHD conduction method. Exp Thermal Fluid Sci 93:108–118

Moatimid GM (1994) Electrohydrodynamic stability with mass and heat transfer of two fluids with a cylindrical interface. Int J Eng Sci 33:125–139

Molki M, Ohadi MM, Bloshteyn M (2000) Frost reduction under intermittent electric field, proceedings of 34" National Heat Transfer Conference, Pittsburgh, PA, NHTC 2000–12052

Motakeff S (1990) Magnetic field elimination of convective interference with segregation during vertical-Bridgman growth of doped semiconductors. J Crystal Growth 104:833–850

Munakata T, Yabe A, Tanasawa I (1993) Effect of electric fields on frosting phenomenon. In: The $6'11$ international symposium on transport phenomena in thermal engineering, pp 381–386

Nasr A, Debbissi C, Orfi J, Nassrallah S (2009) Evaporation of water by natural convection in partially wetted heated vertical plates: effect of the number of the wetted zone. J Eng Appl Sci 4 (1):51–59

Nelson DA, Ohadi MM, Zia S, Whipple RL (1991) Electrostatic effects on heat transfer and pressure drop in cylindrical geometries. Heat Transfer Science and Technology, B.-X. Wang, Ed. Hemisphere, New York, pp 33–39

Ohadi MM, Nelson DA, Zia S (1991a) Heat transfer enhancement of laminar and turbulent pipe flows via corona discharge. Int J Heat Mass Transf 34:1175–1187

Ohadi M, Sharaf N, Nelson D (1991b) Electrohydrodynamic enhancement of heat transfer in a shell-and-tube heat exchanger. Exp Heat Transfer 4(1):19–39

Ohadi MM, Li SS, Dessiatoun S (1994) Electrostatic heat transfer enhancement in a tube bundle gas-to-gas heat exchanger. J Enhanced Heat Transfer 1:327

Paschkewitz JS, Pratt DM (2000) The influence of fluid properties on electrohydrodynamic heat transfer enhancement in liquids under viscous and electrically dominated flow conditions. Exp Thermal Fluid Sci 21:187–197

Pearson MR, Seyed-Yagoobi J (2009) Advances in electrohydrodynamic conduction pumping. IEEE Trans Dielectr Electr Insul 16(2):424–434, 899–910

Sabhapathy P, Salcudean ME (1990) Numerical study of flow and heat transfer in LEC growth of GaAs with an axial magnetic field. J Crystal Growth 104:371–388

Sadek H, Ching CY, Cotton J (2010) Characterization of heat transfer modes of tube side convective condensation under the influence of an applied DC voltage. Int J Heat Mass Transf 53(19–20):4141–4151

Sampath R, Zabaras N (2001) A functional optimization approach to an inverse magneto- convection problem. Comput Methods Appl Mech Eng 190:2063–2097

Satcunanathan S, Deonarine SA (1973) Two-pass solar air heater. Sol Energy 15:41–49

Seyed-Yagoobi J (2005) Electrohydrodynamic pumping of dielectric liquids. J Electrost 63 (6–10):861–869

Terekhov V, Khafaji H, Ekaid A (2015) Numerical simulation for laminar forced convection in a horizontal Insulated Channel with wetted walls, proc. 8th ICCHMT, Istanbul. May:25–28

Terekhov VI, Khafaji HQ, Gorbachev MV (2018) Heat and mass transfer enhancement in laminar forced convection wet channel flows with uniform wall heat flux. Journal of Enhanced Heat Transfer 25(6):565

Trommelmans J, Janssens J, Maelfat F, Berghmans J (1985) Electric field heat transfer augmentation during condensation of nonconducting fluids on a horizontal surface. IEEE Trans Ind Appl IA-21(2):530–534

Van Poppel BP, Desjardins O, Daily JW (2010) A ghost fluid, level set methodology for simulating multiphase electrohydrodynamic flows with application to liquid fuel injection. J Comput Phys 229(20):7977–7996

Wang G, Bao R (2011) Heat transfer augmentation of a transformer oil flow in a smooth tube by EHD effect under high temperatures. J Enhanc Heat Transf 18(2):107–114

Wijeysundera NE, Ah LL, Tjioe LE (1982) Thermal performance study of two-pass solar air heaters. Sol Energy 28:363–370

Yabe A, Taketani T, Kikuchi K, Mori Y, Hijikata K (1987) Augmentation of condensation heat transfer around vertical cooled tubes provided with helical wire electrodes by applying nonuniform electric fields. In: Wang B-X (ed) Heat transfer science and technology. Hemisphere, Washington, DC, pp 812–819

Yabe A, Taketani T, Maki H, Takahashi K, Nakadai Y (1992) Experimental study of electrohydrodynamically(EHD) enhanced evaporator for nonazeotropic mixtures. ASHRAE Trans 98:455–461

Yeh HM, Ho CD, Hou JZ (1999) The improvement of collector efficiency in solar air heaters by simultaneously air flow over and under the absorbing plate. Energy 24:857–871

Yeh HM, Ho CD, Hou JZ (2002) Collector efficiency of double-flow solar air heaters with fins attached. Energy 27:15–727

Chapter 3
Enhancement of Two-Phase Flow Using EHD Technique

3.1 Two-Phase Flow: Condensation

Velkoff and Miller (1965), Yabe (1991) and Sunada et al. (1991) worked with R-113 and R-123 for condensation on a vertical plate. Figure 3.1 shows the results of Sunada et al. (1991). No enhancement was observed until the field strength increased to 4 MV/m. Above 4 MV/m, the character of the condensate film radically changed as shown in Fig. 3.2. Screen electrode gives secondary enhancement known as liquid jetting as shown in Fig. 3.3. Spaced electrodes near the surface cause a jet flow of condensate away from the surface; EHD liquid extraction thins the condensate film. Yabe (1991) combined the liquid jetting and pseudo-dropwise phenomena in condensation on a vertical tube using a helical wire electrode (extraction mode) and a perforated, curved plate (pseudo-dropwise mode). Yamashita et al. (1991) developed a vertical tube condenser. Condensation of perfluorohexane (C_6F_{14}) was considered. Figure 3.4 shows the results of Yamashita et al. (1991).

Holmes and Chapman (1970) worked on the condensation heat transfer of Freon-114. They investigated the non-uniform alternating electric field with voltage ranging between 0 V and 60 kV. Their final conclusion was that the presence of electric filed influences the heat transfer in a positive way and that the heat transfer is 10 times higher than that of without electric field situation.

Vapour space condensation was studied by Wawzyniak and Seyed-Yagoobi (1996). They conducted EHD enhancement studies by using R-113 condensation test on a vertical enhanced tube. Cheung et al. (1999) investigated the effect of electrode-tube gap width using R134a on smooth tubes. Both vertical and horizontal configurations were studied. For both the configurations, the largest gap yielded the highest heat transfer enhancement and with least amount of electrical power to the electrode. Chu et al. (2001) investigated the film condensation enhancement of steam on an integral fin tube. Figure 3.5 shows heat transfer enhancement ratio by EHD versus applied voltage for two different electrode distances (H). The figure shows

© The Author(s), under exclusive license to Springer Nature Switzerland AG 2020
S. K. Saha et al., *Electric Fields, Additives and Simultaneous Heat and Mass Transfer in Heat Transfer Enhancement*, SpringerBriefs in Applied Sciences and Technology, https://doi.org/10.1007/978-3-030-20773-1_3

Fig. 3.1 Heat transfer
enhancement vs. electric
field strength for R-113
condensing at 48 °C on a
vertical plate. From Sunada
et al. (1991)

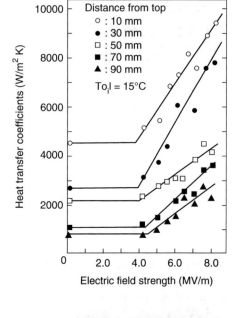

Fig. 3.2 Photograph of
EHD pseudo-dropwise
condensation of R-l 13 at
48.7 °C on a vertical plate
observed by Sunada et al.
(1991). Electric field
strength (7.5 MV/m),
condensing temperature
(48.7 °C). From Sunada
et al. (1991)

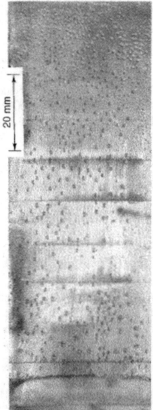

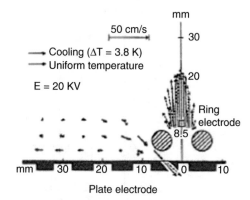

Fig. 3.3 EHD liquid jet phenomenon. From Yabe (1991)

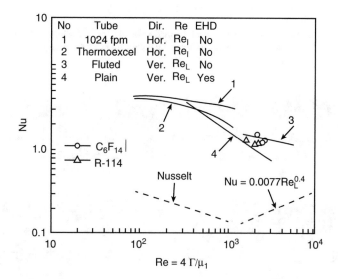

Fig. 3.4 Comparison of test results on 102-tube vertical condenser with test results on single horizontal mechanically enhanced tubes. From Yamashita et al. (1991)

that the enhancement ratio jumps to higher value above a threshold voltage. The threshold voltage is higher for larger H.

Figure 3.6 shows the effect of applied voltage for condensing steam on the integral fin tube studied by Chu et al. (2001). Chu et al. (2001) also presented an analytical model for the prediction of EHD condensation on integral fin tubes. The EHD technique was applied to in-tube condensation by Singh et al. (1997) and Gidwani et al. (2002). Singh et al. (1997) used R-134a for condensation on a conventional smooth tube and on a microfin tube. The enhancement decreased as the mass flux and quality increased. The pressure drop increase was more pronounced than the heat transfer increase. For the microfin tube, the heat transfer enhancement was much less than that of a smooth tube. Gidwani et al. (2002) made

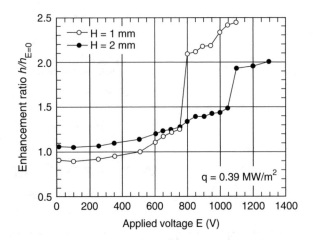

Fig. 3.5 Heat transfer enhancement ratio vs. applied voltage for condensing steam on 16-mm-OD, 1.0-mm integral-fin tube. A 1.0-mm-diameter stainless wire electrode placed under the tube at a distance H = 1 and 2 mm. From Chu et al. (2001)

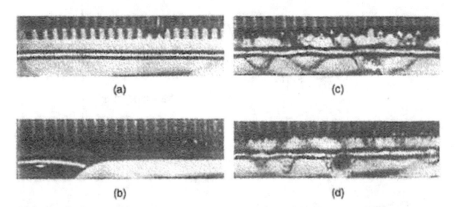

Fig. 3.6 Photographs showing the effect of applied voltage for condensing steam on 16-mm-OD, 1.0-mm integral-fin tube. A 1.0-mm-diameter stainless wire electrode placed under the tube at a distance H = I mm: (**a**) E = 0 V, q = 0 W/m^2, (**b**) E = 0 V, q = 390 W/m^2, (**c**) E = 765 V, q = 390 W/m^2, (**d**) E = 900 V, q = 390 W/m^2. From Chu et al. (2001)

the study on a corrugated tube. The enhancement ratio of the corrugated tube was less than that of the smooth tube.

Kim (2016) studied the condensation heat transfer enhancement and pressure drop of refrigerant R-410A in a smooth tube having an outer diameter 5.00 mm and microfin tubes. He conducted the tests for the mass flow rate range of 50–250 kg/m^{2}s, which is the operational range of residential air conditioners, and fixed heat flux was 4 kW/m^2. Figure 3.7 shows that the heat transfer coefficient increases along the tube length as quality or mass flux increases in a smooth tube. Table 3.1 shows the saturated temperatures, mass fluxes and geometrical parameters used in

Fig. 3.7 The condensation heat transfer coefficient of the 5.00 mm outer diameter smooth tube (Kim 2016)

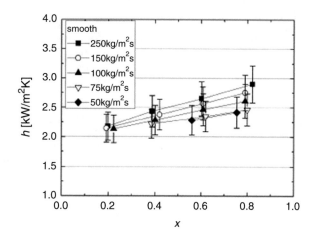

Table 3.1 Available condensation studies in microfin using R-410A (Kim 2016)

Investigators	T_{sat} (°C)	G (kg/m²·s)	Do/D_t (mm)	n	e (mm)	γ (deg)	β (deg)
Kedzierski and Gonclaves (1999)	–	–	9.5/8.51	60	0.2	50	18
			7.94/6.93	50	0.2	57	18
Eckels and Tesene (1999)	40–50	125–500	9.52/8.51	60	0.3	45	18
			7.94/6.93	60	0.3	51	27
Kweon and Kim (2000)	31	97–202	9.52/8.52	60	0.2	53	18
Miyara and Otsubo (2001)	40	100–400	7.0/6.03	50	0.21	41	18
Tang et al. (2000a, b)	40.6	250–810	9.52/8.52	60	0.2	40	18
Goto et al. (2001)	40	200–340	8.0/7.13	55	0.17	55	18
Houfuku et al. (2001)	40	250	7.01/–	50–57	0.15–0.22	15–40	16
							18
Jung et al. (2004)	40	100–300	9.52/8.52	60	0.2	53	18
	20.1–32.0	91–404	–/8.68	60	0.12	48	25
Han and Lee (2005)	19.0–31.7	178–521	–/6.16	60	0.15	53	18
	19.1–32.2	287–921	–/4.84	60	0.13	40	10.3
	18.5–33.8	456–1110	–/3.74	60	0.13	40	9
			9.52/8.60	60	0.12	–	25
			9.52/8.52	60	0.20	53	18
Kim and Shin (2005)	45	180–360	9.52/8.52	60	0.20	40	18
			9.52/8.47	65	0.25	25	15.5
Cavallini et al. (2006)	40	100–800	–/7.69	60	0.23	43	13
			5.0/4.28	38	0.16	40	18
			5.0/4.28	38	0.16	25	18
Zhang et al. (2013)	47	180–650	5.0/4.3	36	0.12	25	18
			5.0/4.3	60	0.12	25	18
			5.0/4,4	52	0.10	20	18
Present study	45	50–250	5.0/4.6	40	0.15	–	10

Table 3.2 Detailed dimensions of the microfin tube and the smooth tube (Kim 2016)

	Microfin tube	Smooth tube
D_o (mm)	5.0	5.0
D_r (mm)	4.6	4.6
D_t (mm)	4.3	4.6
D_m (mm)	4 58	4.6
D_h (mm)	2.99	4.6
A_{fa} (mm^2)	16.5	16.61
A_{ia} (mm^2)	22.34	14.63
A_{im} (mm^2)	14.57	14.61
A_{ia}/A_{im}	1.53	1
P_w (mm)	22.05	14.44
e (mm)	0.15	–
n	40	–
β (deg)	18	–
γ (deg)	40	–

Table 3.3 The heat transfer enhancement factor and pressure drop penalty factor (Kim 2016)

G (kg/m^2·s)	x	EF	PF
	0.82	1.60	1.85
	0.61	1.46	2.14
250	0.40	1.28	1.94
	0.21	1.05	1.90
	0.80	1.55	1.89
	0.60	1.33	1.98
150	0.41	1.10	1.76
	0.21	1.02	2.33
	0.79	1.47	1.87
100	0.60	1.25	1.88
	0.40	1.06	1.92
	0.79	1.68	1.92
75	0.61	1.49	1.70
	0.45	1.38	2.25
50	0.73	1.59	1.43

previous works on condensation heat transfer in microfin tubes using R-410A. Table 3.2 shows the detailed geometrical dimensions of the microfin tube and the smooth tube.

The heat transfer enhancement factor and pressure drop penalty factor were varied with quality and mass flux as shown in the Table 3.3. Eckels and Tesene (1999) reported that enhancement factor increased with increasing of quality at low mass flux of 250 kg/m^{2}s and that it decreased as quality increased at high mass flux of 600 kg/m^{2}s. It was also seen that at a high quality, surface tension induced condensation as well as circumferential redistribution of the thin condensate film increased the heat transfer coefficient. He also measured the variation of frictional pressure drop during the condensation heat transfer tests with mass flux range of

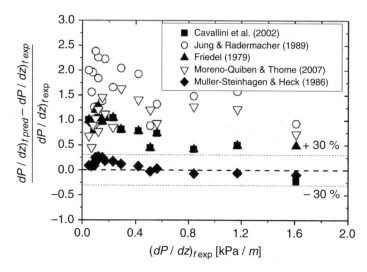

Fig. 3.8 Frictional pressure gradients of the smooth tube compared with the predictions by existing correlations (Kim 2016)

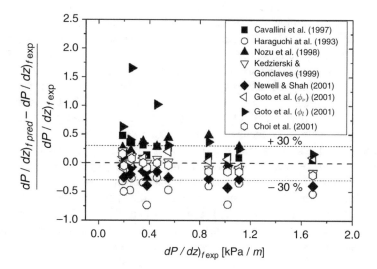

Fig. 3.9 Frictional pressure gradients of the microfin tube compared with the predictions by existing correlations (Kim 2016)

50–250 kg/m²s. Figures 3.8 and 3.9 show the comparison between frictional pressure gradient of smooth tube and microfin tube and predicated the existing correlations, respectively. He observed that in condensation heat transfer, frictional pressure drop of the microfin tube was higher than the smooth tube. Cavallini et al. (1997), Haraguchi et al. (1993), Nozu et al. (1998), Kedzierski and Gonclaves (1999), Newell and Shah (2001), Goto et al. (2001), Choi et al. (2001), Eckels and Tesene

(1999) and Doretti et al. (2013) studied the condensation heat transfer mechanism in a plain tube, enhanced tube, microfin tube, etc.

Shahriari et al. (2017) reviewed the application of electric field on condensation heat transfer by researchers. It included electro-hydrodynamic enhancement of condensation heat transfer, electric field-induced condensation and electric field-enhanced jumping droplet condensation. The EHD-enhanced condensation heat transfer was discovered by Velkoff and Miller (1965) [as per Shahriari et al. (2017)]; there were many progressive works performed by Choi (1968), Holmes and Chapman (1970), Seth and Lee (1974), Didkovsky and Bologa (1981), Bologa et al. (1995) and Cooper (1992). This technique uses high-voltage, low-current AC or DC electric field to obtain and apply force to dielectric liquid condensate.

Butt et al. (2011) claimed that the electric field influenced the vapour pressure. They derived the well-known existing Kelvin equation by using thermodynamic theory and numerical solution-established modified Kelvin equation which included the electric field. Modified equation revealed the reduction in saturation vapour pressure due to electric field and introduces electric field-induced condensation. Further, Kollera and Grigull (1969), Boreyko and Chen (2010), Liu et al. (2014), Boreyko et al. (2011), Boreyko and Collier (2013), Chen et al. (2013), Zhang et al. (2013), Lv et al. (2014), Watson et al. (2015), Wang et al. (2015), Chavez et al. (2016), Boreyko and Chen (2013), Zhang et al. (2015) and Preston et al. (2014) applied jumping motion condensation to different phase-change systems which includes thermal diodes, anti-icing systems, self-cleaning systems, etc. for the study of electric field-enhanced jumping droplet condensation. It occurred when water vapour condenses on super-hydrophobic surface; then the consolidated condensate droplets jumped off spontaneously from the condensing surface in the normal direction. This offers an attractive alternative method for the movement of condensate in the corresponding phase-change system. Its autonomous transportation ability strongly recommends the use in closed-loop systems.

Yabe (1991) discussed that the mechanism of EHD-enhanced condensation consisted of liquid extraction and the formation of small droplets on the condensing surface. Al-Ahmadi and Al-Dadah (2002) provided a comprehensive report on the available correlations for EHD-enhanced condensation heat transfer. Equation 3.1 is the correlation given by Al-Ahmadi and Al-Dadah (2002). Choi and Reynolds' (1965) expression for λ^* (Eq. 3.2) may be used in Eq. 3.1 to evaluate the Nusselt number value by Al-Ahmadi and Al-Dadah (2002) correlation.

$$Nu_e = \frac{h\lambda}{\kappa_i} = A\left[\frac{V^{1.75}h'_{fg}\left(\lambda''/l\right)^{4.5}}{c_{pl}\Delta T}\right]^n \tag{3.1}$$

$$\lambda^{\circ} = \frac{2\pi}{\left(3/4\sigma\right)\left(1-\frac{\varepsilon_g}{\varepsilon_l}\right)^2 \varepsilon_g E_g^2} \tag{3.2}$$

3.2   Two-Phase Flow: Boiling

Falling film evaporation was studied by Yamashita and Yabe (1997). They applied EHD technique to enhance R-123 falling film evaporation from a vertical smooth tube. In addition to the conventional electrode geometries, punched electrodes shown in Fig. 3.10 were tested. Punched electrodes gave higher enhancements compared to the conventional electrodes. In contrast to the condensation case, they observed an optimum applied voltage at which the heat transfer coefficients were maximized. Figure 3.11 shows the optimum Nusselt number versus film Reynolds number obtained with punched electrodes by Yamashita and Yabe (1997). Darabi et al. (2000a) investigated the effect of EHD on falling film evaporation on vertical enhanced tubes using R-134a; the extension of this work was done by Darabi et al. (2000b) to study the horizontal configuration.

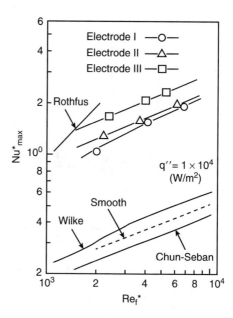

Fig. 3.10 Optimum Nusselt number vs. film Reynolds number, obtained with punched electrodes. From Yamashita and Yabe (1997). Reprinted with permission of ASME

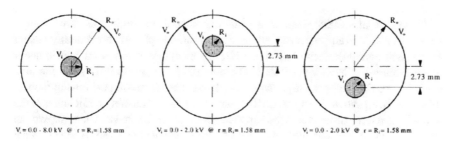

Fig. 3.11 The electrode positions in the test tube (Cotton et al. 2008)

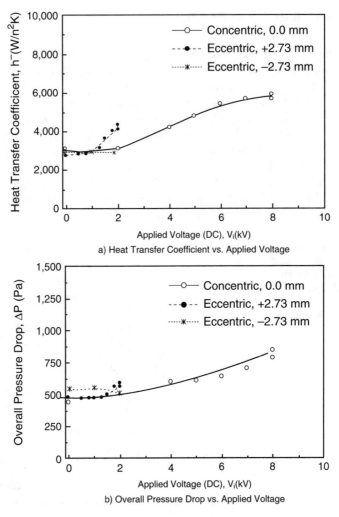

Fig. 3.12 variation of heat transfer coefficient and overall pressure drop with applied voltage (dc) (Cotton et al. 2008)

Cotton et al. (2008) carried out an experiment to study the flow boiling augmentation in an eccentric cylindrical channel placed horizontally using electrohydrodynamic technique. The effect of electrode position on heat transfer augmentation has been investigated. The boiling of refrigerant R-134a on the tube side of a coaxial heat exchanger was considered. The electrode positions in the test tube have been shown in Fig. 3.12. The variation of heat transfer coefficient and the overall pressure drop with the applied voltage (DC) have been shown in Figure 3.13a, b, respectively. Also, the overall enhancement ratio (ratio of heat transfer coefficient at a given voltage to the heat transfer coefficient at no voltage

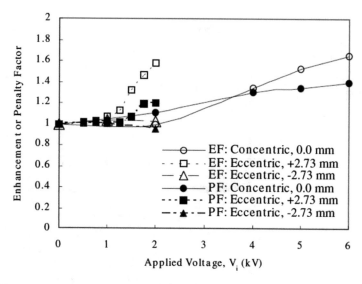

Fig. 3.13 Overall enhancement ratio vs. applied voltage (Cotton et al. 2008)

case) has been shown in Fig. 3.14. For the concentric electrode position, increase in heat transfer coefficient with applied voltage has been observed. Also, the overall enhancement in this case at 6 kV was 1.6 times greater than that compared to no voltage case.

The overall enhancement ratio in case of upper eccentric electrode position was found to be 1.6 at 2 kV applied voltage. It can be noted that in this case the required enhancement ratio can be obtained with three times lesser voltage required in the case of concentric electrode position while the same voltage applied in the case of lower eccentric electrode position resulted in negligible enhancement. From the experimental results, they concluded that the electrode placed eccentrically with +2.73 eccentricity showed 60% increment in heat transfer coefficient and only 1.2 times increase in pressure drop at 2 kV applied voltage. Also, the increase in heat transfer enhancement with electric field strength was observed while the opposite trend has been reported for heat transfer coefficient variation with mass flux and inlet quality. The heat transfer enhancement in case of electrode placed eccentrically with −2.73 eccentricity for inlet quality less than 10% was almost negligible.

Moradian and Saidi (2008) presented the effect of electric field on the bubble behaviour. They presented a theoretical model for electrohydrodynamically enhanced boiling heat transfer. The schematic representation of the effect of electric field on spherical bubble under uniform electric field and non-uniform electric field has been shown in Fig. 3.15. The application of external electric field to film boiling destabilizes the liquid film for heat transfer enhancement. The heat transfer enhancement mechanisms in nucleate boiling are the electro-convection and effects on bubble dynamics. These mechanisms are dependent on the intensity of electric field and relaxation time. The relaxation time may be defined as the time required

Fig. 3.14 Schematic representation of electric field effect on a spherical bubble: (**a**) uniform electric field and (**b**) non-uniform electric field (Moradian and Saidi 2008)

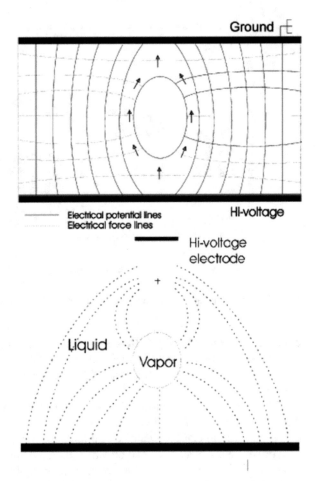

for the electric force to affect the bubble generation. It is explicit that the effect of electric field on boiling heat transfer would be negligible when the relaxation time is greater than the bubble generation frequency. Then the heat transfer enhancement is by electro-convection mechanism.

Falling film evaporation on single tube and tube bundles was investigated by Thome (2017). He investigated the influence of film feed rate, tube layout, heat flux, enhancement geometry effects, tube layouts, tube pitch, nozzle type and height on the performance of heat transfer. They found that enhanced tube evaporator with horizontal bundles required lesser refrigerant charge than flooded evaporator for falling film evaporation. Zeng et al. (1997, 1998) extended their test on single tube spray evaporation with pure ammonia in a plain horizontal stainless tube and compared the results with falling film performance and again extended their test on falling film evaporation of ammonia on low carbon steel fin tube and corrugated tube. Yamashita and Yabe (1997) investigated the falling film evaporation with R-123 on vertical stainless steel tube diameter of 25.4 mm by using

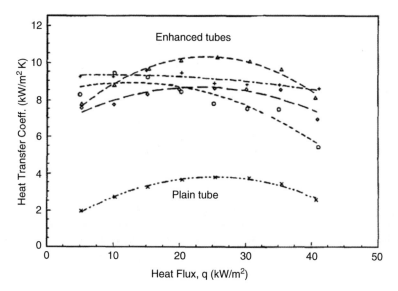
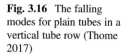

Fig. 3.15 Falling film heat transfer coefficients for R-134a on single enhanced tubes and plain tube [o-Turbo-B, Δ-Turbo-Cii, ◇-Wieland-SE, + Wieland-SC, ×-Plain] (Thome 2017)

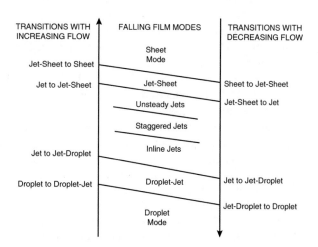

Fig. 3.16 The falling modes for plain tubes in a vertical tube row (Thome 2017)

electrohydrodynamic technique. Chyu et al. (1982) studied the falling film heat transfer coefficient for water on three types of tube: (a) a horizontal plain tube, (b) a high flux porous coated tube and (c) a Gewa-T tube. They found that Gewa-T geometry promoted enhanced thin film evaporation without producing nucleate boiling. The Turbo-Cii gave the best performance.

Figure 3.16 shows a comparison of heat transfer coefficient of the complex geometry tubes to the plain tube for pure R-134a at 2 °C. Moeykens and Pate (1995) investigated the effect of nozzle height and orifice diameter for spray

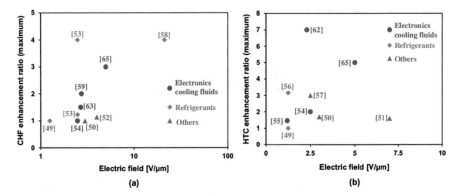

Fig. 3.17 Enhancement of (**a**) critical heat flux and (**b**) heat transfer coefficient vs. the applied electric field for insulating liquids. Experimental data were obtained from the citations in brackets. The baseline for the enhancement was the heat transfer in the absence of an electric field for that particular study (Shahriari et al. 2017)

evaporation with 20 low finned tubes in a triangular tube for R-134a at 2 °C. Ulucakli (1996) measured falling film coefficient for subcooled water without evaporation and investigated the local film effect in wavy laminar film flow. He found that waves profoundly influenced the heat transfer. Alhusseini et al. (1996) investigated the effect of the critical heat flux (CHF) on nucleate boiling of evaporating falling films on outside of vertical plain tube. They found that CHF increased with increasing film velocity. Figure 3.17 depicts their schematic summary of falling film modes and transitions for plain tube. Hu and Jacobi (1996a, b), Cerza (1992) and Nuinrich (1996) conducted an experimental work on falling film evaporation in heat transfer enhancement.

Shahriari et al. (2017) reviewed augmentation of boiling and condensation heat transfer with the application of electric field. They reviewed electric field-based controlled heat transfer using electrically insulating liquid and later with electrically conducting liquids. The electrically insulating (dielectric) liquids (refrigerants and electronic cooling fluid) were used by Siedel et al. (2011), Chen et al. (2013), Liu et al. (2006), Zhang et al. (2010), Takata et al. (2003), Schweizer et al. (2013), di Marco et al. (2003), Quan et al. (2015), Sheikhbahai et al. (2012), Bologa et al. (2012), Wang et al. (2009), Zaghdoudi and Lallemand (2005), di Marco and Grassi (2011), Kweon and Kim (2000), Zaghdoudi and Lallemand (1999), Carrica et al. (1997), Mahmoudi et al. (2014), Migliaccio and Garimella (2013), Takano et al. (1996), McGranaghan and Robinson (2014), Verplaetsen and Berghmans (1999) and Pandey et al. (2016) for boiling heat transfer enhancements. Similarly, electric field effect on bubble dynamics was investigated by Siedel et al. (2011) and Liu et al. (2006), and they studied parameters like bubble growth, detachment, departure velocity with trajectory, etc. Shahriari et al. (2017) presented Fig. 3.18 indicating the impact of boiling heat transfer in maximum enhancement of CHF and heat transfer coefficient graphs with corresponding electric field.

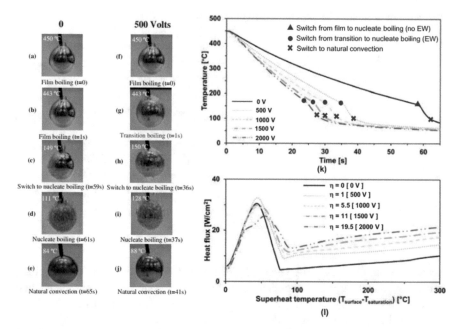

Fig. 3.18 Boiling patterns on (**a**)–(**e**) non-electro-wetted and (**f**)–(**j**) electro-wetted copper spheres. The vapor film (Leidenfrost state) is (**g**) immediately disrupted in the electro-wetted sphere and (**h**) collapses completely in 36 s. (**j**) Natural convection starts at 41 s. For the non-electro-wetted sphere, (**c**) a stable Leidenfrost state persists until 59 s, and (**e**) natural convection starts at 65 s. (**k**) Electrically tunable cooling curves showing voltage-dependent cooling, with modified boiling regimes; points where boiling regimes change are marked on the curves. (**l**) Heat dissipation capacity vs. superheat for various electro-wetting voltages can be extracted from the cooling curve (Shahriari et al. 2017)

Shahriari et al. (2017) elaborated the results of experiment and numerical simulation carried out to study the effect of electric field on boiling heat transfer of conducting liquids. The typical conducting liquids used were water, methanol, ethanol and other organic solvents having conductivities from 10^{-6} to 10^{-4} S/m. Cho et al. (2015) used water for performing nucleate boiling under the influence of electric field. They applied electric potential less than 2 V for controlling bubble nucleation, and they demonstrated the application of voltage in turning on and off bubbles spatially and temporally. This optimization of spatial configuration in turn enhances heat transfer. Cho et al. (2015) investigated the influence of surfactants and electric field and showed that they were advantageous in improving boiling heat transfer. Takano et al. (1996) claimed that electric field across vapour gap eliminated the Leiden frost state, and it augmented heat transfer by 2.6–2.8 times with R-113 and ethanol working fluids, respectively.

Similarly, Shahriari et al. (2017) presented the suppression of Leiden frost state using organic solvents for deionized water and liquids at the surface temperature 550 °C. Shahriari et al. (2017) investigated rapid cooling due to the expulsion of film boiling during quenching process. The cooling curve was found to be lowered

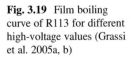

Fig. 3.19 Film boiling curve of R113 for different high-voltage values (Grassi et al. 2005a, b)

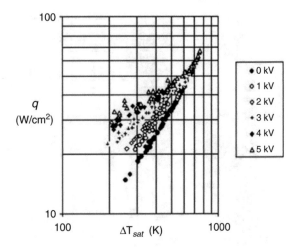

because of the significant influence of Leidenfrost effect. So, they studied quenching process with and without the application of electric field. They used isopropanol bath for quenching of copper and stainless steel spheres (1 inch diameter) starting from 800 °C. They presented Fig. 3.19 showing the presence and absence of 500 V DC signal associated with quenching. It was observed that Leidenfrost layer developed instantly in the absence of DC signal. The electrical voltages replaced film boiling with the transition boiling followed by nucleate boiling and then convection. They concluded that high voltage pushed heat transfer although it got saturated after a limiting high voltage.

Grassi and Testi (2006) presented the experimental results related to the pool film boiling heat transfer with testing on wires at atmospheric pressure. They used R-113 ($\varepsilon_r = 2.41$) and Vertrel XF ($\varepsilon_r = 6.72$) for testing electric permittivity in this process. They obtained increased heat flux at a given wall superheat and presented it in Figs. 3.20 and 3.21. Between the two fluids, the Vertrel XF (polar fluid) performed better than weakly polar R-113 liquid. They concluded that oscillation wavelength and wall superheat were strongly interconnected. The transition between regimes revealed that wall superheat overpowers the heat flux, and the transition from one-dimensional to a two-dimensional regime was detected.

EHD significantly enhances the nucleate boiling. Figure 3.22 shows the results of Yabe (1991) on a plain horizontal tube using a mix of R-11 and ethanol. The performance is compared to that of high-flux porous surfaces, and the Thermoexcel-E mechanically enhanced surface geometries both operating without EHD enhancement. The performance increases with increasing applied voltage, and it is as high as or sometimes even higher than that provided by the mechanically enhanced surfaces. The proposed mechanism for EHD enhancement on a plain surface works as follows:

The EHD body force pushes the vapour bubbles away from the electrode, and the greatest force occurs on the part of the bubble closest to the electrode. The bubble is pushed towards the heating surface (away from the electrode), the radial components

Fig. 3.20 Film boiling curve of Vertrel XF for different high-voltage values (Grassi et al. 2005a, b)

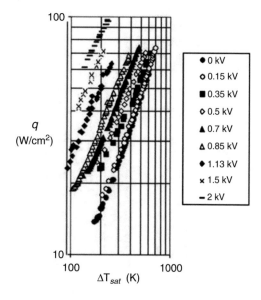

Fig. 3.21 Yabe's (1991) test results for EHD boiling enhancement on a plain horizontal tube using a 90/10% mixture (by weight) of R-114 and ethanol. Also shown are data for boiling on mechanically enhanced tubes without EHD enhancement

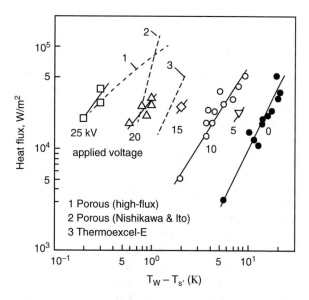

of the body force cause violent movement of the bubbles on the heating surface. This enhances thin film evaporation at the base of the bubble and promotes the breakup of large bubbles into a greater number of smaller diameter bubbles. The observation of Ogata and Yabe (1991) reaffirmed the above analysis, and they observed the effect of an electric field on a single air bubble in a silicone/ethanol mixture. The bubble breakup was explained by the Taylor instability in the electric field on the liquid–gas interface, whose shape is deformed by the field.

Fig. 3.22 R-114 boiling at
21.5 °C on a 1428 fins/m
horizontal, integral-fin tube
tested by Cooper (1990).
From Cooper (1990)

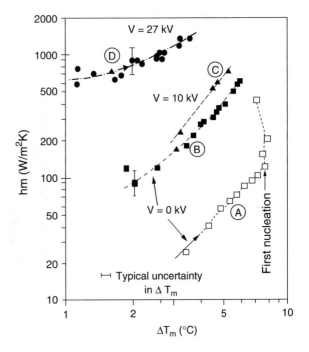

Boiling mixtures have a smaller heat transfer coefficient than a mono (pure) fluid. Yabe (1991) conjectured that violent mixing at the heated surface should be beneficial for the enhancement of mixture boiling. Figure 3.22 corroborates this theory, and this mechanism is also supported by visualization studies of Ohadi et al. (1992). Ohadi et al. (1992) boiled oil–refrigerant mixtures using R-11 and R-123. High-speed digital camera captured the data which showed that EHD force causes the number of bubbles to increase and causes the diameter of the bubbles to decrease. Kweon and Kim (2000) examined the bubble dynamics by measurements while working with R-113 pool boiling on a single wire, and their observations support the previous experience. The nucleation site density and the bubble frequency increased as the electric field density increased. The bubble diameter, however, decreased as the electric field density increased.

Figure 3.23 shows the results of Cooper (1990) on a standard integral fin tube using R-114. Cooper (1992) argues that EHD forces activate nucleation sites at lower super heat than are required without an electric field. After activation, the forces provided by the electric field further enhance nucleate boiling. The EHD application eliminates the boiling hysteresis: this has been observed by Cooper (1990, 1992), Oh and Kwak (2000), Zaghdoudi and Lallemand (2005). Zaghdoudi and Lallemand (2005) observed that the electric field modifies the surface tension and the fluid does not wet the surface. The above results have been corroborated by Damianidis et al. (1992), who observed that electric field increases the boiling coefficient.

Papar et al. (1993) investigated the effect of electrode geometry on EHD-enhanced boiling of R-123/oil mixtures on a horizontal smooth tube. Singh et al. (1995a) extended this study to integral fin tubes. A mesh electrode was used. Yan et al. (1996) investigated the EHD effect on the pool boiling of high-performance boiling tubes. Darabi et al. (2000c) provided additional data on R-134a boiling of integral fin tubes. Zaghdoudi and Lallemand (1999) investigated the effect of electric field polarity on nuclear pool boiling of n-pentane on a horizontal plane. The negative polarity configuration yielded a larger enhancement than the positive polarity configuration.

Yabe et al. (1992) studied convective vaporization of R-123 and R-134a inside a circular tube. They observed good enhancement. However, the enhancement ratio decreased as the mass flux increased. Better enhancement is observed in case of nuclear pool boiling. Singh et al. (1994, 1995b) experimented with R-123 and R-134a with the application of EHD in a circular tube. Bryan and Seyed-Yagoobi (2001) conducted R-134a boiling in a smooth tube. The electric force causes an early dryout because of the extraction of liquid from the wall since the liquid film is thin. Norris et al. (1999) reported the reduction of heat transfer coefficient at high qualities by EHD.

Ogata et al. (1992) experimented with R-123 boiling on a horizontal bundle of 50 plain tubes using six axial wire electrodes spaced at 3.0 mm from each tube. The performance was competitive with that of a single tube. Cheung et al. (1995) conducted EHD-enhanced boiling of R-134a in a bundle of seven integral fin tubes. The enhancement decreased as the heat flux increased. Karayiannis (1998) investigated the EHD effect for R-123 and R-11 boiling on a horizontal tube bundle of five tubes. The enhancement with R-11 was only marginal because of longer electric relaxation time of R-11 than the bubble detachment period. If the relaxation time is long, the electron forces generated on the bubbles are too weak to affect the bubble behaviour significantly.

Zaghdoudi and Lallemand (2005) investigated the effect of EHD on CHF from a horizontal plate. Yabe (1991) reported 20% increase of CHF of R-113 when uniform electric field of 20 kV/cm is applied.

Cooper (1990) extended the Rohsenow (1952) model and developed a correlation for EHD-enhanced pool boiling heat transfer, given in Eq. 3.3, where all the coefficients and indices may be obtained from his paper.

$$h_E = ah_oNe^{0.165}Re_o^b \qquad (3.3)$$

References

Al-Ahmadi A, Al-Dadah RK (2002) A new set of correlations for EHD condensation heat transfer of tubular systems. Appl Thermal Eng 22:1981–2001

Alhusseini AA, Hoke BC, Chen JC (1996) Critical heat flux in falling films undergoing nucleate boiling. In Chen JC (eds) Convective flow boiling. Taylor Francis: 339–344 (Proc. of Convective Flow Boiling Conj: in Banff, Canada, April 30–May 5, 1995)

Bologa MK, Korovkin VP, Savin IK (1995) Mechanism of condensation heat transfer enhancement in an electric field and the role of capillary processes. Int J Heat Mass Transf 38(1):175–182

Bologa MK, Kozhevnikov IV, Mardarskii OI, Polikarpov AA (2012) Boiling heat transfer in the field of electric forces. Surf Eng Appl Electrochem 48(4):329–331

Boreyko JB, Chen CH (2010) Self-propelled jumping drops on superhydrophobic surfaces. Phys Fluids 22(9):091110

Boreyko JB, Chen CH (2013) Vapor chambers with jumping-drop liquid return from super-hydrophobic condensers. Int J Heat Mass Transf 61:409–418

Boreyko JB, Collier CP (2013) Delayed frost growth on jumping-drop superhydrophobic surfaces. ACS Nano 7(2):1618–1627

Boreyko JB, Zhao Y, Chen CH (2011) Planar jumping-drop thermal diodes. Appl Phys Lett 99 (23):234105

Bryan JE, Seyed-Yagoobi Y (2001) Influence of flow regime, heat flux, and mass flux on electrohydrodynamically enhanced convective boiling. J Heat Transfer 123:355–367

Butt HJ, Untch MB, Golriz A, Pihan SA, Berger R (2011) Electric-field-induced condensation: an extension of the kelvin equation. Phys Rev E 83(6):061604

Carrica P, Di Marco P, Grassi W (1997) Nucleate pool boiling in the presence of an electric field: effect of subcooling and heat-up rate. Exp Thermal Fluid Sci 15(3):213–220

Cavallini A, Del Col D, Doretti L, Longo GA, Rossetto L (1997) Pressure drop during condensation and vaporization of refrigerants inside enhanced tubes. Heat Technol 15(1):3–10

Cavallini A, Del Col D, Mancin S, Rossetto L (2006) Thermal performance of R-410A condensing in a microfin tube. In: Proc Int Refrig Conf R178

Cerza M (1992) Nucleate boiling in thin falling liquid films. Pool and External Flow Boiling Conference, Santa Barbara: 459–466

Chavez RL, Liu F, Feng JJ, Chen CH (2016) Capillary-inertial colloidal catapults upon drop coalescence. Appl Phys Lett 109(1):011601

Chen X, Ma R, Zhou H, Zhou X, Che L, Yao S, Wang Z (2013) Activating the microscale edge effect in a hierarchical surface for frosting suppression and defrosting promotion. Sci Rep 3:2515

Cheung K, Ohadi MM, Dessiatoun SV (1999) EHD-assistecl external condensation of R-134a on smooth horizontal and vertical tubes. Int J Heat Mass Transf 42:1747–1755

Cheung KH, Ohadi MM, Dessiatoun S (1995) Compound enhancement of boiling heat transfer of R-134a in a tube bundle. ASHRAE Trans Symp 101(Part 1):1009–1019

Cho HJ, Mizerak JP, Wang EN (2015) Turning bubbles on and off during boiling using charged surfactants. Nature Communications 6:8599

Choi HY (1968) Electrohydrodynamic condensation heat transfer. J Heat Transf 90(1):98–102

Choi JY, Kedzierski MA, Domanski PA (2001) Generalized pressure drop correlation for evapo-ration and condensation in smooth and microfin tubes, proc. of IIF-IIR commission B1, Paderborn, Germany, B4: 9–16

Chu RC, Nishio S, Tanasawa I (2001) Enhancement of condensation heat transfer on a finned condensation. Proc Third ASMEIJSME Thermal Eng Conf 3:47–53

Chyu MC, Bergles AE, Mayinger F (1982) Enhancement of horizontal tube spray film evaporators, Proc. 7th international heat transfer con/, Munich, 6, 275–280

Cooper P (1990) EHD enhancement of nucleate boiling. J Heat Transfer 112:458–464

Cooper P (1992) Practical design aspects of EHD heat transfer enhancement in evaporators. In ASHRAE Winter Meeting, Anaheim. CA, USA, 01/25–29/92: 445–454

Cotton JS, Robinson AJ, Chang JS, Shoukri M (2008) Electrohydrodynamic enhancement of flow boiling in an eccentric horizontal cylindrical channel. J Enhanc Heat Transf 15(3):183–198

Damianidis C, Karayinnis T, Al-Dadah RK, James RW, Collins MW, Allen PHG (1992) EHD boiling enhancement in shell-and-tube evaporators and its application to refrigeration plants. ASH RAE Trans 98(Part 2):462–473

Darabi J, Ohadi MM, Dessiatoun SV (2000a) Falling film and spray evaporation enhancement using an applied electric field. J Heat Transfer 122:741–748

Darabi J, Ohadi MM, Dessiatoun SV (2000b) Augmentation of thin falling-film evaporation on horizontal tubes using an applied electric field. J Heat Transfer 122:391–398

Darabi J, Ohadi MM, Dessiatoun SV (2000c) Compound augmentation of pool boiling on three selected commercial tubes. J Enhanced Heat Transfer 7:347

Di Marco P, Grassi W (2011) Effects of external electric field on pool boiling: comparison of terrestrial and microgravity data in the ARIEL experiment. Exp Thermal Fluid Sci 35 (5):780–787

Di Marco P, Grassi W, Memoli G, Takamasa T, Tomiyama A, Hosokawa S (2003) Influence of electric field on single gas-bubble growth and detachment in microgravity. Int J Multiphase Flow 29(4):559–578

Didkovsky AB, Bologa MK (1981) Vapour film condensation heat transfer and hydrodynamics under the influence of an electric field. Int J Heat Mass Transf 24(5):811–819

Doretti L, Zilio C, Mancin S, Cavallini A (2013) Condensation flow patterns inside plain and microfin tubes: a review. Int J Refrig 36:567–587

Eckels SJ, Tesene BA (1999) A comparison of R-22, R-134A, R-410A and R-407C condensation performance in smooth and enhanced tubes, part I: heat transfer. ASHRAE Trans 105:428–441

Gidwani A, Molki M, Ohadi MM (2002) EHD-enhanced condensation of alternative refrigerants in smooth and corrugated tubes. Int J HVAC & R Res 8(3):219–238

Goto M, Inoue N, Ishiwatari N (2001) Condensation and evaporation heat transfer of R-410A inside internally grooved horizontal tubes. Int J Refrig 24:628–638

Grassi W, Testi D, Saputelli M (2005a) EHD enhanced heat transfer in a vertical annulus. Int Commun Heat Mass Transf 32(6):748–757

Grassi W, Testi D, Saputelli M (2005b) Heat transfer enhancement in a vertical annulus by electrophoretic forces acting on a dielectric liquid. Int J Therm Sci 44(11):1072–1077

Grassi W, Testi D (2006) Heat transfer augmentation by ion injection in an annular duct. J Heat Transfer — Trans. ASME. 128:283–289

Han D, Lee KJ (2005) Experimental study on condensation heat transfer enhancement and pressure drop penalty factors in four microfin tubes. Int J Heat Mass Transfer 48:3804–3816

Haraguchi H, Koyama S, Esaki J, Fujii T (1993) Condensation heat transfer of refrigerants Hcfc134A, Hcfc123 and Hcfc22 in a horizontal smooth tube and a horizontal microfin tube, in proc. of 30th Natl. Symp. of Japan, Yokohama 343–345

Holmes RE, Chapman AJ (1970) Condensation of Freon-114 in the presence of a strong nounniform alternating electric field. J Heat Transf Trans ASME 92:616–620

Houfuku M, Suzuki Y, Inui K (2001) High-performance, lightweight THERMOFIN tubes for air conditioners using alternative refrigerants. Hitachi Cable Review 2001:97–100

Hu X, Jacobi AM (1996a) The Intertube falling film: part I-flow characteristics, mode transitions, and hysteresis. J Heat Transf 118:616–625

Hu X, Jacobi AM (1996b) The Intertube falling film: part 2-mode effects on sensible heat transfer to a falling liquid film. J Heat Transf 118:626–633

Jung D, An K, Park J (2004) Nucleate boiling heat transfer coefficients of HCFC22, HFC134a, HFC125, and HFC32 on various enhanced tubes. Int J Refrigeration 27:202–206

Karayiannis TG (1998) EHD boiling heat transfer enhancement of R 123 and Rl 1 on a tube bundle, of R-123 and their enhancement using the EHD technique. J Enhan Heat Transfer 2:209

Kedzierski MA, Gonclaves JM (1999) Horizontal convective condensation of alternative refrigerants within a microfin tube. J Enhanc Heat Transf 6:161–178

Kim NH (2016) Condensation heat transfer and pressure drop of R-410a in 5.0-mm-od smooth or microfin tubes at low mass fluxes. J Enhanc Heat Transf 23(5):120–145

Kim MH, Shin JS (2005) Condensation heat transfer of R-22 and R-410A in horizontal smooth and microfin tubes. Int J Refrig 28:949–957

Kollera M, Grigull U (1969) The bouncing off phenomenon of droplets with condensation of mercury. Heat Mass Transf 2(1):31–35

Kweon YC, Kim MH (2000) Experimental study on nucleate boiling enhancement and bubble dynamic behavior in saturated pool boiling using a non-uniform dc electric field. Int J Multiphase Flow 26(8):1351–1368

Liu F, Ghigliotti G, Feng JJ, Chen CH (2014) Numerical simulations of self-propelled jumping upon drop coalescence on non-wetting surfaces. J Fluid mechanics 752:39–65

Liu Z, Herman C, Mewes D (2006) Visualization of bubble detachment and coalescence under the influence of a nonuniform electric field. Exp Thermal Fluid Sci 31(2):151–163

Lv J, Song Y, Jiang L, Wang J (2014) Bio-inspired strategies for anti-icing. ACS Nano 8 (4):3152–3169

Mahmoudi SR, Adamiak K, Castle GP (2014) Flattening of boiling curves at post-CHF regime in the presence of localized electrostatic fields. Int J Heat Mass Transf 68:203–210

McGranaghan GJ, Robinson AJ (2014) The mechanisms of heat transfer during convective boiling under the influence of AC electric fields. Int J Heat Mass Transf 73:376–388

Migliaccio CP, Garimella SV (2013) Evaporative heat transfer from an electro-wetted liquid ribbon on a heated substrate. Int J Heat Mass Transf 57(1):73–81

Miyara A, Otsubo Y (2001) Condensation heat transfer of herringbone micro fin tubes. Exp Heat Trans. Fluid mechanics, and thermodynamics 2001, Edzioni ETS, Pisa, Italy, pp 381-386

Moeykens SA, Pate MB (1995) The effects of nozzle height and orifice Siz.E on spray evaporation heat transfer performance for a low-finned, triangular-pitch tube bundle with R-134a. ASHRAE Trans 101(2):420–433

Moradian A, Saidi MS (2008) Electrohydrodynamically enhanced nucleation phenomenon: a theoretical study. J Enhanc Heat Transf 15(1):1–15

Newell TA, Shah RK (2001) An assessment of refrigerant heat transfer, pressure drop and void fraction effects in microfin tubes. Int J HVAC&R Res 7(2):125–153

Norris CE, Cotton JS, Shoukri M, Chang J-S, Smith-Pollard T (1999) Electrohydrodynamic effects on flow redistribution and convective boiling in horizontal concentric tubes under high inlet quality conditions. ASHRAE Trans 105(Part 1):222–236

Nozu S, Katayama H, Nakata H, Honda H (1998) Condensation of refrigerant Cfc11 in horizontal microfin tubes (proposal of a correlation equation for frictional pressure gradient). Exp Thermal Fluid Sci 18:82–96

Nuinrich R (1996) Falling film evaporation of soluble mixtures. Convective Flow Boiling 335:125–135

Ogata J, Iwafuji Y, Shimada Y, Yamaziki T (1992) Boiling heat transfer enhancement in tubebundle evaporator utilizing electric field effects. ASHRAE Trans 98(Part 2):435–444

Oh S-D, Kwak HY (2000) A study of bubble behavior and boiling heat transfer enhancement under electric field. Heat Transfer Eng 21(4):33–45

Ohadi M, Faani M, Papar R, Radermacher R, Ng T (1992) EHD heat transfer enhancement of shell-side boiling heat transfer coefficients of R-123/oil mixture. ASHRAE Trans 98(Part 2):424–434

Pandey V, Biswas G, Dalal A (2016) Effect of superheat and electric field on saturated film boiling. Physics of Fluids 28(5):052102

Papar RA, Ohadi MM, Kumar A, Ansari AI (1993) Effect of electrode geometry on EHD enhanced boiling of R-123/oil mixture. ASHRAE Trans 99(Part 1):1237–1243

Preston DJ, Miljkovic N, Enright R, Wang EN (2014) ΔV Ē. J Heat Transf 136:080909–080901

Quan X, Gao M, Cheng P, Li J (2015) An experimental investigation of pool boiling heat transfer on smooth/rib surfaces under an electric field. Int J Heat Mass Transf 85:595–608

Rohsenow WM (1952) A method of correlating heat transfer data for surface boiling of liquids. ASME Trans 74:969–976

Schweizer N, Di Marco P, Stephan P (2013) Investigation of wall temperature and heat flux distribution during nucleate boiling in the presence of an electric field and in variable gravity. Exp Thermal Fluid Sci 44:419–430

Seth AK, Lee L (1974) The effect of an electric field in the presence of noncondensable gas on film condensation heat transfer. J Heat Transf 96(2):257–258

Shahriari A, Birbarah P, Oh J, Miljkovic N, Bahadur V (2017) Electric field–based control and enhancement of boiling and condensation. Nanosc Microsc Thermo-Phys Eng 21(2):102–121

Sheikhbahai M, Esfahany MN, Etesami N (2012) Experimental investigation of pool boiling of Fe3O4/ethylene glycol–water nanofluid in electric field. Int J Therm Sci 62:149–153

Siedel S, Cioulachtjian S, Robinson AJ, Bonjour J (2011) Electric field effects during nucleate boiling from an artificial nucleation site. Exp Thermal Fluid Sci 35(5):762–771

Singh A, Ohadi M, Dessiatoun S (1997) EHD enhancement of in-tube condensation heat transfer of alternate refrigerant, R-134a. ASHRAE Trans 103(1):97–100

Singh A, Ohadi MM, Dessiatoun S (1995) EHD-enhanced boiling of R-123 over commercially available enhanced tubes. J Heat Transfer 117:1070–1073

Singh A, Ohadi MM, Dessiatoun S, Chu W (1994) In-tube boiling heat transfer enhancement of R-123 using the EHD technique. ASHRAE Trans 100(Part 2):818–825

Sunada K, Yabe A, Taketani T, Yoshizawa Y (1991) Experimental study of EHD pseudo-dropwise condensation. Proc ASME/JSME Therm Eng 3:61–67

Tang L, Ohadi MM, Johnson AT (2000a) Flow condensation in smooth and micro-fin tubes with HCFC-22, HFC-134a and HFC-410 refrigerants part 11: design equations. J Enhan Heat Trans 7(5):311–326

Tang L, Ohadi MM, Johnson AT (2000b) Flow condensation in smooth and micro-fin tubes with HCFC-22, HFC-134a and HFC-410A refrigerants part 1: experimental results. J Enhan Heat Trans 7(5):289–310

Takano K, Tanasawa I, Nishio S (1996) Enhancement of evaporation of a liquid droplet using EHD effect: criteria for instability of gas-liquid interface under electric field. J Enhan Heat Trans 3(1):72

Takata Y, Shirakawa H, Tanaka K, Ito T (2003) Numerical study on motion of a single bubble exerted by non-uniform electric field. Int J Transp Phenom 5:247–258

Ulucakli E (1996) Heat transfer in a subcooled falling liquid film. In Chen JC (eds) Convective flow boiling. Taylor & Francis pp. 329-334 (proc. c,f convective flow boiling Conf in ban Canada, April 30–May 5, 1995)

Velkoff HR, Miller JH (1965) Condensation of vapor on a vertical plate with a transverse electrostatic field. J Heat Transf 87(2):197–201

Verplaetsen FM, Berghmans JA (1999) Film boiling of an electrically insulating fluid in the presence of an electric field. Heat Mass Transf 35(3):235–241

Wang P, Lewin PL, Swaffield DJ, Chen G (2009) Electric field effects on boiling heat transfer of liquid nitrogen. Cryogenics 49(8):379–389

Wang Q, Yao X, Liu H, Quéré D, Jiang L (2015) Self-removal of condensed water on the legs of water striders. Proc Natl Acad Sci 112(30):9247–9252

Watson GS, Schwarzkopf L, Cribb BW, Myhra S, Gellender M, Watson JA (2015) Removal mechanisms of dew via self-propulsion off the gecko skin. J R Soc Interface 12(105):20141396

Wawzyniak M, Seyed-Yagoobi J (1996) Experimental study of electrohydrodynamically augmented condensation heat transfer on a smooth and an enhanced tube. J Heat Transfer 118:499–501

Yabe A (1991) Active heat transfer enhancement by applying electric fields. In: Proc. third ASME/JSME thermal Eng. C01~f. 3, xv-xxiii

Yamashita K, Kumagai M, Sekita S, Yabe A, Taketani T, Kikuchi K (1991) Heat transfer characteristics of an EHD condenser. Proc. Third ASME/JSME Joint Thermal Eng. Conj, Reno, Nevada, pp 61–67

Yamashita K, Yabe A (1997) Electrohydrodynamic enhancement of falling film evaporation heat transfer and its long-term effect on heat exchangers. J. Heat Transfer 119(2):339–347

Yan YY, Neve RS, Allen PHG (1996) EHD effects on nucleate boiling at passively enhanced surfaces. Exp Heat Transfer 9(3):195–212

Zaghdoudi MC, Lallemand M (1999) Analysis of the polarity influence on nucleate pool boiling under a DC electric field. J Heat Transf 121(4):856–864

Zaghdoudi MC, Lallemand M (2005) Pool boiling heat transfer enhancement by means of high DC electric field. Arab J Sci Eng 30(2):112–125

Zeng X, Chyu M-C, Ayub ZH (1997) Performance of nozzle-sprayed Ammonia evaporator with square-pitch plain-tube bundle. ASH RAE Trans 03(2), Paper):4059

Zeng X, Chyu M-C, Ayub ZH (1998) Ammonia spray evaporation heat transfer performance of single low-fin and corrugated tubes. ASHRAE Trans 104(1): Paper SF-98-15-2 (4109)

Zhang K, Liu F, Williams AJ, Qu X, Feng JJ, Chen CH (2015) Self-propelled droplet removal from hydrophobic fiber-based coalescers. Phys Rev Lett 115(7):074502

Zhang Q, He M, Chen J, Wang J, Song Y, Jiang L (2013) Anti-icing surfaces based on enhanced self-propelled jumping of condensed water microdroplets. Chem Commun 49(40):4516–4518

Zhang HB, Yan YY, Zu YQ (2010) Numerical modelling of EHD effects on heat transfer and bubble shapes of nucleate boiling. Appl Math Model 34(3):626–638

Chapter 4
Mass Transfer in the Gas Phase

Either condensation into a liquid film or evaporation from a liquid film is an example of mass transfer in the gas phase. In most of the cases, the inert component is air, although it may be any inert component. We will now look into the case of condensation with noncondensable gases. Figure 4.1 shows the cross section of a tube wall with a condensing vapour on one side and cooling water on the other side. The solid curves in the condensing region show the temperature and pressure distributions with noncondensable gas, whereas the broken lines indicate conditions that would exist if a noncondensable gas was not present. The curves in the figure are self-explanatory as regards the temperature of the bulk vapour and the liquid–vapour interface where condensation takes place. The difference between the interface temperature and the saturation temperature reduces because (a) condensing vapour exists at its partial pressure and the saturation temperature reduces and (b) condensing vapour must diffuse through the vapour–gas mixture and the vapour partial pressure at the interface is less than that of the bulk mixture. This reduces the saturation temperature at the interface. High vapour velocity or enhancement will reduce the temperature drop across the gas boundary layer.

The heat transfer mechanism in dropwise condensation using noncondensable gases was studied by Ma et al. (2007). They carried out experiments on dropwise, filmwise and combined dropwise-filmwise condensation with and without using noncondensable gases. The condensation of steam–air mixture on vertical plates with special designs was considered. The applications of condensation using noncondensable gases in different areas such as seawater distillation, sorption refrigeration, latent heat recovery from flue vapour and separation of mixtures of non-azeotropic organic vapours. The schematic representation of dropwise condensation and filmwise condensation of steam–air mixture on vertical plates has been shown in Fig. 4.2. The dropwise, filmwise and combined dropwise-filmwise condensation patterns have been shown in Fig. 4.3.

The surface area for condensation was observed to change with the mode of condensation. The entire surface of droplets can be considered as the condensation

© The Author(s), under exclusive license to Springer Nature Switzerland AG 2020
S. K. Saha et al., *Electric Fields, Additives and Simultaneous Heat and Mass Transfer in Heat Transfer Enhancement*, SpringerBriefs in Applied Sciences and Technology, https://doi.org/10.1007/978-3-030-20773-1_4

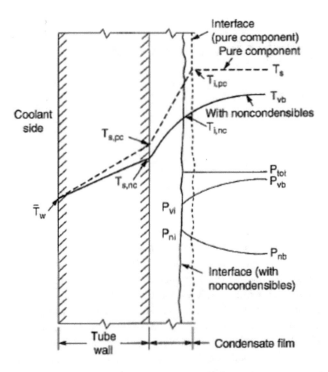

Fig. 4.1 Temperature and pressure profiles for the condensation of pure vapour (solid lines) and with noncondensable gas (dashed line). From Webb et al. (1980)

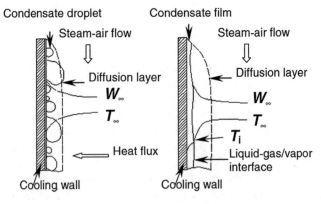

Fig. 4.2 Schematic representation of dropwise condensation and filmwise condensation of steam–air mixture on vertical plates (Ma et al. 2007)

area. The maximum surface area was observed in the case of dropwise condensation. The increase in surface area in the case of dropwise condensation and combined dropwise-filmwise condensation was due to increase in droplet surface. The heat flux and heat transfer coefficients versus degree of saturation for the experiments done in

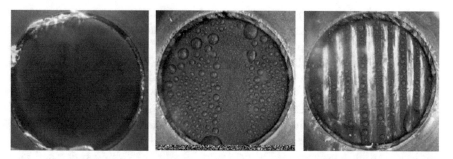

Fig. 4.3 Dropwise, filmwise and combined dropwise-filmwise condensation patterns (Ma et al. 2007)

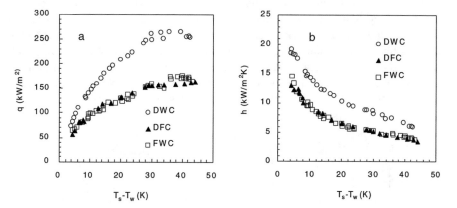

Fig. 4.4 Heat flux and heat transfer coefficients vs. degree of saturation in the presence of noncondensable gases (Ma et al. 2007)

the presence of noncondensable gases have been shown in Fig. 4.4. They observed that although half of the condensation surface is covered by droplets in case of dropwise-filmwise condensation, there was no effect on heat and mass transfer. Increase in heat and mass transfer by 30–80% has been observed in the case of dropwise condensation. This is because, the small droplets grow till the critical size is gained and after that they fall back on to the surface. Then disturbance is created due to surging, pumping and shearing on boundary layer, and heat and mass transfer was augmented. They concluded that organic coating yielded augmented heat transfer for dropwise condensation with noncondensable gases.

Huang and Deng (2018) worked on heat and mass transfer enhancement of falling liquid films subjected to evaporative and sensible heating in the vertical tube. The improvement in performance of evaporators is advantageous for energy conservation, process design, chemical industry, food industry, seawater desalting and waste reduction. Falling film evaporation in horizontal tubes was investigated by Tian et al. (2012), Li et al. (2011) and Zhao et al. (2018). However, vertical tube performance was examined by few researchers. The existing heat transfer enhancement techniques are tube gas ventilation, tube insertion and tube shape modification. Luan et al. (1933) and Sun and Lin (1993) investigated tube gas ventilation and found that it was inefficient for inner parts and it had limited effect because gases disturbed film surfaces only. Huang et al. (2006), Duryodhan et al. (2015) and Pehlivan (2013) used converging-diverging tube and achieved high performance in the enhancement of heat and mass transfer. Wang et al. (2007) used the converging-diverging tube for single-phase convective heat transfer.

As the vertical tube falling film evaporation was little explored, Huang and Deng (2018) examined the vertical converging-diverging tube for falling film evaporation for heat transfer improvement. They compared it with the built-in spring coil tube and a smooth tube for performance evaluation. They used deionized water, and the liquid films on tube wall had uniform thickness and stable state for achieving uniform even film spreading. Otherwise, local drying-out of tube walls may occur.

The falling film evaporation experimentation was performed by altering water flow rate, and evaporative heat transfer coefficient was analysed. They used two converging-diverging tubes for testing and for comparing; a smooth and spring coil tube was considered. The schematic representation of the tubes has shown in Fig. 4.5, and their dimensions were presented in Table 4.1. They measured liquid mass at initial stage and after testing, vapour mass generated by evaporation, mass

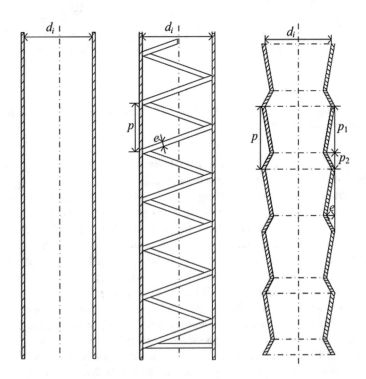

Fig. 4.5 Schematic representation of smooth tube, built-in spring coil tube and converging-diverging tube (Huang and Deng 2018)

Table 4.1 Main structure dimension of heat transfer tubes (Huang and Deng 2018)

Tube shape	Outer diameter, d_o (m)	Inner diameter, d_i (m)	Length of analysis segment, L (m)	Pitch spacing/ node spacing, p (m)	Length of converging segment, p_1 (m)	Length of diverging segment, p_2 (m)	Rib height, e (m)
Converging-diverging tube 1	0.019	0.016	2.3	0.014	0.0035	0.0105	0.002
Converging-diverging tube 2	0.019	0.016	2.3	0.014	0.0105	0.0035	0.002
Smooth tube	0.019	0.017	2.3	–	–	–	–
Spring coil	0.017	0.015	2.3	0.0035	–	–	0.001

and temperature of the remaining unevaporated liquid and liquid temperature for the calculation of results. The whole evaporation process can be modelled into two stages: the first stage for sensible heating and the second stage for evaporation. They calculated the overall heat transfer coefficient of falling film evaporation (K) by Eq. (4.1) where T_k is the outside temperature of the tube and T belongs to liquid film temperature. The steam condensation heat transfer coefficient (h_o) was evaluated by the Nusselt filmwise condensation correlation for outside of the tube. They calculated the overall heat transfer coefficient for the falling film evaporation section (K_2) by Eq. (4.2).

$$K = \frac{q}{T_k - T} \qquad (4.1)$$

$$K_2 = \frac{1}{{}^{1}/_{h_o} + \left({}^{d_o}/_{2\lambda_s}\right) \ln \left(d_o / d_i\right) + \left({}^{d_o}/_{d_i h_2}\right)} \qquad (4.2)$$

Due to axial symmetry, they simplified the converging-diverging tube to two-dimensional outer surface with some vertical tilt.

They concluded that errors related to mass and heat transfer calculations were within ±3% and ±8%, respectively. They used empirical formulae of Wilke and Herrmann (1962). They presented the effect of mass flow rate in Fig. 4.6. They investigated in the Reynolds number range of 1000–2500 in the four tubes and found that as Reynolds number increased, the heat transfer coefficients related to built-in spring coil tube and converging-diverging tubes (C-D tube) were proportionally increased. The plot clearly summarized that the heat transfer coefficient of smooth tube overpowers that of the others in the Reynolds number range of 1000–1500. According to heat transfer coefficient, the performance in descending order was observed for smooth tube followed by built-in spring coil tube and then converging-diverging tube. However, this was changed in the Reynolds number range of 1800–2500. They observed that converging-diverging tube performance was best,

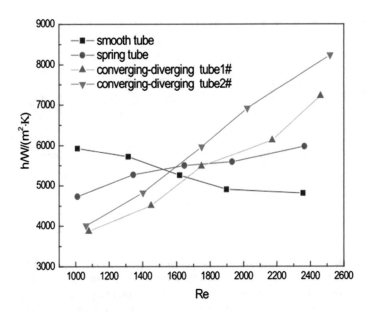

Fig. 4.6 Relationship between the evaporation heat transfer coefficient and the liquid film Reynolds number (Huang and Deng 2018)

followed by built-in spring coil tubes and at last smooth tube. This suggests that converging-diverging effects are profound for higher Reynolds number.

The valid reason behind this situation can be stated as average liquid film thickness was thinner in C-D tube than that of smooth tube at low Reynolds number. It reduced with the increase in Reynolds number. Higher Reynolds number intensified the turbulence in C-D tube, and turbulence-induced heat transfer overcomes the film thickness leading to increase in heat transfer coefficient. They calculated heat transfer coefficients at Reynolds number 2356 for convergent-divergent tubes 1 and 2 which were 1.14 and 1.3 times higher than that of built-in spring coil tube and 1.42 and 1.62 times higher than that of the smooth tube, respectively. Similar trends were observed for perimeter flow where smooth tube shows the downtrend. They found that evaporation heat transfer coefficient of converging-diverging tubes 1 and 2 and built-in spring coil tube were 32.6%, 29.7% and 25.2% larger than that of corresponding sensible heat transfer coefficients, respectively. This conclusion was obtained from Fig. 4.7.

The heat transfer coefficients of falling films subjected to sensible heating were plotted against liquid film Reynolds number (700–1700) in Fig. 4.8. The plotted graph revealed that sensible heat transfer coefficient was higher in converging-diverging tube than that in built-in spring coil tube followed by smooth tube in the Reynolds number 700–1100. This order changed for Reynolds number 1200–1700. Here, heat transfer coefficient for converging-diverging tube was superior to that for built-in spring coil tube and smooth tube. The sensible heat transfer values for converging-diverging tube 1 and 2 were 1.16 and 1.28 times higher than that for

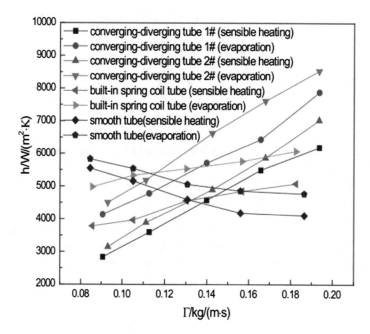

Fig. 4.7 Comparison of sensible heating and evaporation heat transfer coefficients of falling films in the four tubes (Huang and Deng 2018)

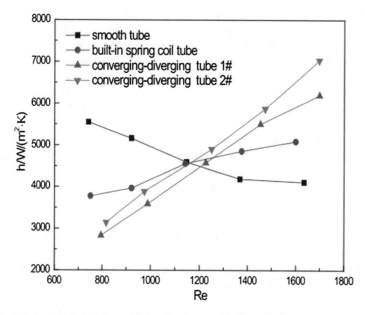

Fig. 4.8 Effect of liquid film Reynolds number on sensible heat transfer coefficient (Huang and Deng 2018)

Table 4.2 Associations of falling film sensible heating and evaporation within four tubes (Huang and Deng 2018)

Tube shape	Sensible heating	Evaporative heating
Converging-diverging tubes	$h^+ = 0.000072 \, Re^{1.0646}\text{Pr}^{1/3}$	$h^+ = 0.000386 \, Re^{0.7941}\text{Pr}^{1/3}$
Built-in spring coil tube	$h^+ = 0.0047 \, Re^{0.476}\text{Pr}^{1/3}$	$h^+ = 0.0161 \, Re^{0.292}\text{Pr}^{1/3}$
Smooth tube	$h^+ = 2.5135 \, Re^{0.412}\text{Pr}^{1/3}$	$h^+ = 1.0117 \, Re^{-0.273}\text{Pr}^{1/3}$

built-in spring coil and 1.44 and 1.58 times higher than smooth tube, respectively. They listed sensible heating and evaporation heat transfer coefficient correlations in Table 4.2.

The heat transfer rate from the vapour mixture to the interface is given by Eq. 4.3 in which the first term is for the sensible heat transfer from the vapour and the second term is the latent heat transfer. In Eq. 4.3, the Ackermann correction factor accounts for the effect of mass flux on the sensible heat transfer rate. The gas phase mass transfer coefficient may be obtained from the heat and mass transfer analogy (Eq. 4.4), where the various terms are represented by standard symbols.

$$q = h_g \alpha_s (T_{vb} - T_i) + K_p i_{gv}(p_{av} - p_{vi}) \qquad (4.3)$$

$$\frac{h_g}{c_p} \text{Pr}^{2/3} = K_p p M Sc^{\frac{2}{3}} \qquad (4.4)$$

Techniques applicable to the enhancement of the gas phase heat transfer coefficient will also be effective in enhancing the mass transfer coefficient. Enhancement is required at the gas–liquid interface, rather than at the pipe wall, and the selection of the enhancement should account for the effect of the liquid film thickness. The two-dimensional rib roughness on the tube outer surface increases the gas phase heat transfer coefficient and hence the mass transfer coefficient. Any effective enhancement technique for convective heat transfer to gases should be beneficial to convective condensation with noncondensable gases. Webb (1991) discussed the heat–mass transfer analogy, and the same nomenclature used by him is used in this research monograph.

The important evaporation process from a water film into an airstream is used in several engineering applications, such as cooling tower, humidifier, evaporative fluid coolers and evaporative condensers. Heat is transferred from a water film by evaporation into a flowing airstream. The mass transfer impedance is in the gas boundary layer. Enhancement may be made of the gas phase mass transfer coefficient.

Cooling tower performance may be improved by cooling tower packing. Water evaporated at the liquid film surface is transferred by convective heat and mass transfer to the moist air flowing over the surface. Equations 4.5, 4.6 and 4.7 (Webb and Kim 2005; Webb 1988, 1991) are the relevant equations in this regard.

$$q = h_g \alpha_s (T_{vb} - T_i) + K_w i_{gv}(W_b - W_i) \qquad (4.5)$$

Fig. 4.9 Transverse-rib concept applied to enhance mass transfer in the vapour boundary layer. From and Perez-Blanco (1986)

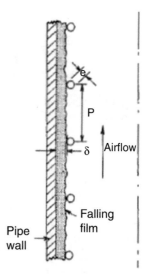

$$K_W = pM(1-y)^2 K_p = \frac{h_g}{c_p}(1-y)^2 \left(\frac{Pr}{Sc}\right)^{\frac{2}{3}} \tag{4.6}$$

$$q = K_W(i_b - i_i) \tag{4.7}$$

Webb and Perez-Blanco (1986) worked with evaporation of water into an air–steam mixture flowing inside a vertical tube (Fig. 4.9), and this was a gas phase enhancement process. Figure 4.9 shows the transverse rib concept applied to enhance mass transfer in the vapour boundary layer. A water film was draining down the inside surface of a round vertical tube, with moist air in the counterflow. The roughness was displaced from the wall by a distance equal to the calculated liquid film thickness, and the ribs were not contained in the liquid film. Webb et al. (1980) gave the correlation for calculating the heat transfer coefficient of the rib roughness.

Gu et al. (2018) numerically studied the heat and mass transfer characteristics between the surface of water and airstream. The hot air flowing over the water surface results in heat transfer by convection, mass transfer and evaporation of water into the airstream. Their objective was to study the effect of mass flux and variation in air properties with temperature on heat and mass transfer. The phenomenon of evaporation of water into airstream involves combined heat and mass transfer. It is commonly encountered in applications such as drying, concentration and desalination of water. The prediction of heat and mass transfer coefficients in such process is generally quite difficult and complex. Thus, for low mass fluxes, the heat transfer coefficients can be evaluated using Colburn–Chilton analogy which cannot be used in case of higher mass fluxes.

The process of convective heat transfer and mass transfer between the surface of water and airstream has been shown in Fig. 4.10. The combined heat and mass

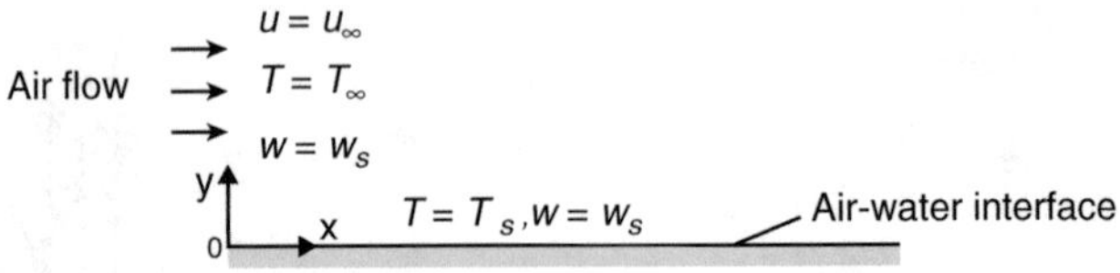

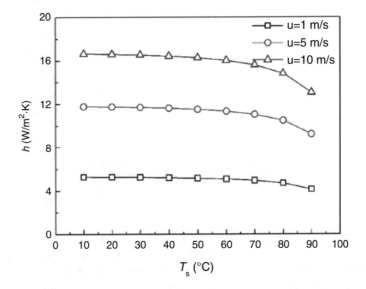

Fig. 4.10 Process of convective heat transfer and mass transfer between surface of water and airstream (Gu et al. 2018)

Fig. 4.11 Variations of overall heat transfer coefficient with water surface temperature for different air velocities (Gu et al. 2018)

transfer is observed due to zero specific humidity of free airstream and high temperature difference between water surface and free airstream. In order to simplify the complexity of the problem, few assumptions were made. The airflow is a steady laminar boundary layer flow. The water surface is assumed to be stationary, having uniform temperature. The specific humidity of the saturated air at the interface can be calculated at the water surface temperature. The variation of heat flux and mass flux with water surface temperature has been shown in Figs. 4.11 and 4.12, respectively. Figures 4.13 and 4.14 show the Nusselt number and Sherwood number variation with Reynolds number, respectively.

They observed that the mass flux increased with increase in water surface temperature and resulted in increased heat and mass transfer coefficients. The effect of mass flux on heat and mass transfer coefficients was evaluated by calculating the correction factors. The correction factors for heat and mass transfer coefficients were noted to decrease with increasing mass flux. The effect of air velocity on the correction factors was negligible. They reported that the Colburn–Chilton analogy

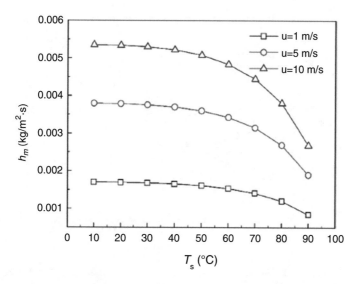

Fig. 4.12 Variations of overall mass transfer coefficient with water surface temperature for different air velocities (Gu et al. 2018)

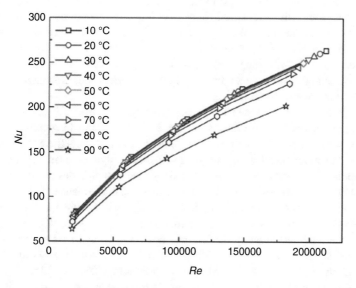

Fig. 4.13 Variations of Nusselt number with Reynolds number for different water surface temperatures (Gu et al. 2018)

to evaluate heat transfer coefficients was applicable for low mass flux, and the deviation of numerical results from those of Colburn–Chilton analogy was only 5% for water surface temperatures less than 60 °C. The variation of air properties showed a clear and complex effect on heat transfer coefficient. On the other hand, the effect of air property variation on mass transfer coefficient was unobvious.

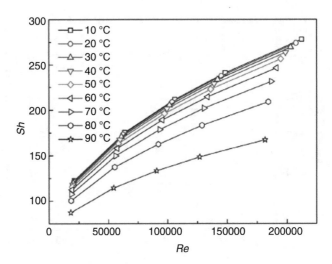

Fig. 4.14 Variations of Sherwood number with Reynolds number for different water surface temperatures (Gu et al. 2018)

Yoshida and Hyodo (1970), Chow and Chung (1983), Yuan et al. (2004), Tang and Etzion (2004), Boukadida and Nasrallah (2001), Stegou-Sagia (1996), Talukdar et al. (2008), Raimundo et al. (2014), Poós and Varju (2017), Iskra and Simonson (2007), Wei et al. (2017), Kumar et al. (2014), Schwartze and Brocker (2000), Volchkov et al. (2007), Jang et al. (2005), Wan et al. (2017a, b) and Azizi et al. (2007) presented similar works on evaporation of water into airstream with combined heat transfer and mass transfer.

Dehumidification of finned tube heat exchangers is used for cooling air and moisture condensates on the finned surface, if the fin surface temperature is below the dew point. A surface geometry with high heat transfer coefficient should also enhance the mass transfer coefficient and provide high performance under wet conditions. Moisture condensation on the finned surface provides a naturally occurring enhancement. Senshu et al. (1981) measured the performance of convex louvre fin geometry under dehumidifying condition for which moisture condensation occurred on the finned surface. The predicted mass transfer coefficient based on specific humidity or enthalpy driving potential was compared by them with the experimental value. Senshu et al. (1981) developed an equation for this purpose. The heat–mass transfer analogy showed that the draining condensate did not bridge the fin louvres or substantially alter the airflow pattern over the louvres. This theory assumes the surface to be fully wetted, which is not always true.

Lee et al. (2016a) presented correlations of Nu and Sh for a desiccant cooling technology using LiCl solution and moist air in plate-type dehumidifier. A hydrophilic coated plate with grooves provided to improve the wettability was considered. The dehumidification process is of four types: a) dehumidification by cooling, b) dehumidification by compression, c) dehumidification by adsorption and d) dehumidification by absorption. The process of dehumidification by cooling is the

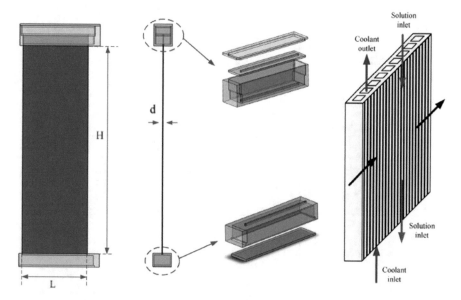

Fig. 4.15 Schematic representation of the plate-type dehumidifying heat exchanger (Lee et al. 2016a)

popularly used method for air-conditioning systems. Although better cooling performance is obtained using this method, whenever the temperature of the solution drops, reheating is required which makes the process quite expensive. Also, at lower dew point, the condensate freezes blocking the airflow.

On the other hand, the dehumidification process by compression involves consumption of large amount of power. The dehumidification process using adsorption can be used for applications with low dew points, but its thermal effectiveness is less than that of an absorption dehumidification process. The adsorption dehumidification process uses a solid adsorbent while absorption dehumidification process uses liquid in contact with air. For improved contact with air, a liquid desiccant may be used which is suitable for dehumidifier having large size. Low vapour pressure, low viscosity, chemical stability and non-corrosiveness are some of the desirable properties of liquid desiccant. The schematic representation of the plate-type dehumidifying heat exchanger has been shown in Fig. 4.15. The experimental conditions used for the present study and similar research works have been presented in Table 4.3.

The rate of absorption versus the air velocity has been shown in Fig. 4.16. The increase in absorption rate with air velocity was observed due to improved mass transfer. However, after the velocity of 0.95 m/s, the absorption rate was observed to decrease with air velocity. This can be attributed by the decrease in contact time at higher velocities. The results obtained from the correlations developed for Nu and Sh have been presented in Figs. 4.17, 4.18, 4.19 and 4.20 for moist air side and liquid solution side.

Table 4.3 Specifications and experimental conditions for the present study and references (Lee et al. 2016a)

References	Desiccant	Flow type	Dehumidifier	Air mass flow rate (kg/s)	Temperature of air (°C)	Solution mass flow rate (kg/s)	Concentration (%)	Temperature of solution (°C)	Specific surface area (m²/m³)
Present	LiCl	Cross flow	Plate	0.0012–0.00739	35	00005–0.0015	0.4–045	43	
Liu et al. (2007)	LiBr	Cross flow	Celdek structured packing	0.427	29.1–299		0.43–0435	21.4–22.4	397
				0.31–047	39.9–30.3	0.0103	0.428–0.431		
				0.31	29.9–32.7	0.0103	0.43–048	21	
Yang et al. (2015)	LiCl	Cross flow	Ultrasonic atomization (nozzle)	0.0189–0.03194	25	00033–0.0133	026–0.38	25	
Koronaki et al. (2013)	LiCl, LiBr, CaCl₂	Counter flow	Packed column	0.00642–0.107		0.00653–0.0653	0.3–0.42		233
Moon et al. (2009)	CaCl₂	Cross flow	Structured packed tower	0.092–0.1985	30	0.048–0.289	0.32–0.42	30	608
Dinner et al.(2014)	CaCl₂	Cross flow	Polypropylene-parallel plate	0.16	30	001–0.06	0.4	34–42	80
Bassuoni (2011)	CaCl₂	Cross flow	Structured packing-corrugation angle of 60°	0.084–0.144	31	0.01–0.06	–	–	390
Yin et al. (2015)	LiCl	Counter flow	Packed bed	0.002–0.0068	25.9–29	0.0086–0.0172	0.34–0.4	28.3–29.5	500
Gao et al. (2012)	LiCl	Cross flow	Celdek packing	0.08–0.14	33	01–0.26	0.38	38	396
Zhang (2011)	LiCl	Counter flow	Hollow membrane	0.0015–0.0057	35	0.002747	0.35	25	750

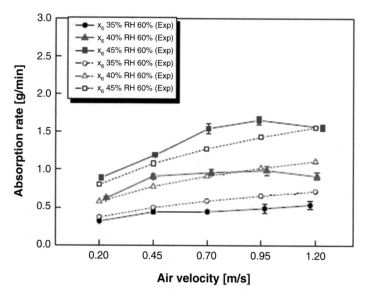

Fig. 4.16 Rate of absorption vs. the air velocity (Lee et al. 2016a)

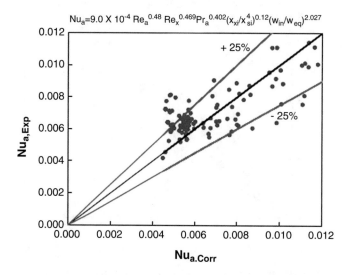

Fig. 4.17 The experimental correlation of Nu for moist air side (Lee et al. 2016a)

They concluded that there was 11.35–66.33% increase in absorption rate with increasing liquid desiccant flow and air velocity, respectively. The effect of air velocity on absorption rate was more profound than that of liquid desiccant flow. Also, higher absorption rates were observed at higher concentration of liquid desiccant and higher relative humidity. This increase is mainly due to the increase in partial pressure at air–liquid desiccant interface at higher relative humidity and

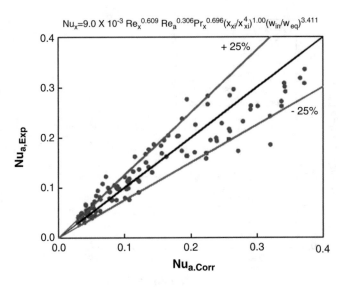

Fig. 4.18 The experimental correlation of Nu for liquid solution side (Lee et al. 2016a)

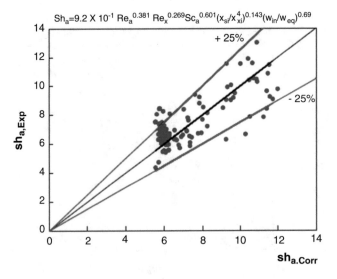

Fig. 4.19 The experimental correlation of Sh for moist air side (Lee et al. 2016a)

liquid desiccant concentration. They developed correlations for Nu and Sh for air side and liquid desiccant side with an error of 25% which can be used for plate-type dehumidifiers with lithium chloride solution.

Similar works have been presented by Jung et al. (2014), Lee et al. (2016b), Fumo and Goswami (2002), Rahamah et al. (1998), Chung et al. (1996), Liu et al. (2007), Zhang et al. (2010), Yang et al. (2015), Moon et al. (2009), Gao et al. (2012) and Zhang (2011).

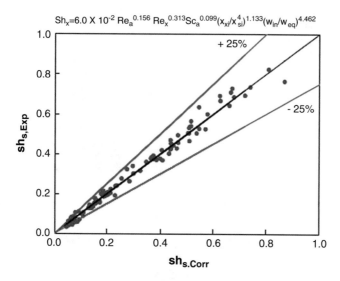

Fig. 4.20 The experimental correlation of Sh for liquid solution side (Lee et al. 2016a)

Wang et al. (1997) found j and f factors of finned tube heat exchangers under dehumidifying conditions. They tested nine staggered plain fin tube heat exchangers. They observed that the friction factors of wet coils were much greater than those of the dry coils. Similar observations were made for fin and tube heat exchangers having louvre fins, Wang et al. (2000). The wet surface momentum loss and pressure drop can be significantly reduced by applying a hydrophilic coating on the fin surface. Water film enhancement of finned tube exchanger was dealt by Yang and Clark (1975). They sprayed a water mist on the face of an automotive radiator. Only 10% performance increase was observed. Evaporation from a wetted surface into the airstream is also very important. The water flows as a film on the fin surface, with water evaporation from the film surface. Good surface wetting is extremely important for the enhancement technique to be effective. Qualification of the degree of surface wetting and drainage characteristics of the water film are important.

Zhang et al. (2019) studied the simultaneous heat and mass transfer processes taking place in an absorption system and presented the irreversibility analysis. The irreversibility analysis depicts the scope for improving the process efficiency. Although the entransy dissipation concept exists for absorption systems, the entransy related to combined heat and mass transfer is not available in literature. Zhang et al. (2019) have thus defined the concentration entransy in their work. The definition of concentration entransy is based on saturation temperature of the absorbent. He explained the need for characterization of irreversibilities in terms of transport phenomena rather than heat–work conversion. They have presented models for entransy dissipation calculation in absorber.

The heat and mass flow for a differential element in falling film has been shown in Fig. 4.21, where mdx is the change in flow rate of the solution, dm_v is the vapour

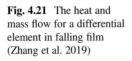

Fig. 4.21 The heat and mass flow for a differential element in falling film (Zhang et al. 2019)

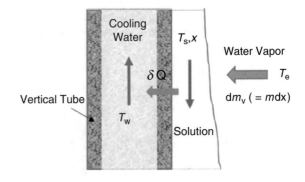

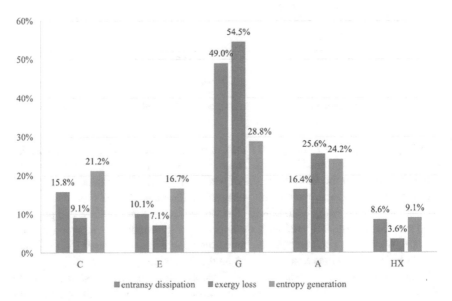

Fig. 4.22 Entransy dissipation, energy loss and entropy generation proportions for each component (Zhang et al. 2019)

absorption rate and δQ is the heat transfer to cooling water. The entransy dissipation (kW/K), energy loss (kW) and entropy generation (kW/K) in the absorption system have been evaluated. Figure 4.22 shows the component-wise entransy dissipation, energy loss and entropy generation. They observed that the dimensions of the irreversibilities associated with mass transfer process were same as those of irreversibilities associated with heat transfer process. They presented new T-Q diagrams which help in understanding whether it is the flow ratio or the heat and mass transfer coefficients which need adjustments or improvements.

Grossman (1983), Mittermaier et al. (2014) and Meyer and Ziegler (2014) studied the simultaneous heat and mass transfer in film absorption under laminar flow regime. Combined heat and mass transfer in falling film and bubble mode absorbers has been presented by Abid et al. (2012). More relevant works on this topic can be

Table 4.4 Geometries tested by Hudina and Sommer (1988)

Geometry code	F_S	F_0
Tube diameter (d_o)	17.3	17.3
Transverse tube pitch (S_t)	60	57
Longitudinal tube pitch (S_t)	25	45
Number of tube rows	6	6
Fin pitch	0.028	0.028
Fin thickness	0.30	0.30

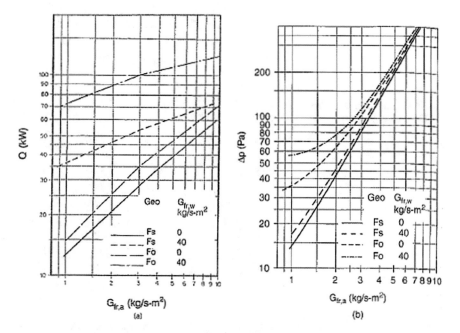

Fig. 4.23 Test results of Hudina and Sommer (1988) for spray enhancement of the Table 16.1 finned tube heat exchangers. (**a**) Heat transfer enhancement, (**b**) pressure drop enhancement. From Hudina and Sommer (1988)

obtained from Park et al. (2003), Hoffmann et al. (1996), Kim et al. (2003) and Kulankara and Herold (2002).

Hudina and Sommer (1988) used mist coating to enhance two different plate-finned tube heat exchangers having plain fins (Table 4.4). Figures 4.23 and 4.24 show the results of Hudina and Sommer (1988), who worked with fixed water spray rate, tube side flow rate, 50% relative humidity, 293 K water spray temperature and inlet air temperatures between 298 and 303 K. Figure 4.24 shows the effect of water spray rate on the heat transfer enhancement level. The enhancement level increased with increasing water spray rate.

Sommer (1984) observed that a uniform thin film does not exist over the fin surface. At low temperature difference between the entering fluid streams (ITD), Sommer observed that the spray water is collected on the fins near the air inlet in the

Fig. 4.24 Effect of water spray rate on heat transfer enhancement for Fs finned tube geometry of Table 16.1 tested by Hudina and Sommer (1988). From Hudina and Sommer (1988)

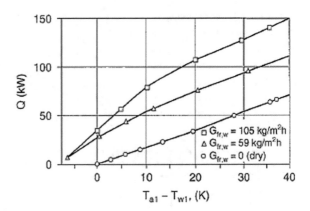

Table 4.5 Geometries Tested by Kreid et al. (1983b)

	Geometry		
	A	B	C
$A_o/A_n N$ (m²/m²)	13.88	27.05	14.70
A_c/A_{fr} (m²/m²)	0.495	0.534	0.509
S_r (mm)	25.0	38.0	30.93
S_t (mm)	60.0	38.0	35.7
D_h (mm)	3.87	2.96	3.18
d_o (mm)	18.0	16.0	16.0
t_f (mm)	0.43	0.22	0.20
Fins/m	354	393	315
N	6	3	4
d_t (mm)	20.2	13.4	13.4

form of large droplets and eventually the fin spacing is blocked. The water may pass through the heat exchanger as fine entrained droplets. High contact angle prevents wetting of the fin surface by water.

Kreid et al. (1983b) investigated heat exchangers oriented with the fins in the vertical direction to facilitate water drainage. There was deluge of water from a supply manifold above the heat exchanger onto the heat exchanger. Table 4.5 shows the geometries tested by Kreid et al. (1983b). Kreid et al. (1978, 1983a) developed an analytical model which predicted the results under wet conditions. The heat transfer performance of a finned tube heat exchanger can be significantly increased with a good mechanism for wetting. Johnson et al. (1981) developed a water-enhanced dry cooling tower. Bentley et al. (1978) tested a vertical plate having spaced, gravity-drained water channels. They observed that, for a certain specific range of parameters, the heat transfer by evaporation varied between 20 and 40% of the total heat transfer. They used repeated-rib roughness. They did not address practical means to supply and distribute the water to finned tube banks. Figure 4.25 shows surface A with interrupted strip fins which prevent lateral spreading of a water film entering the front in a spray. Also, in Fig. 4.25, the concave channel elements of surface B wavy

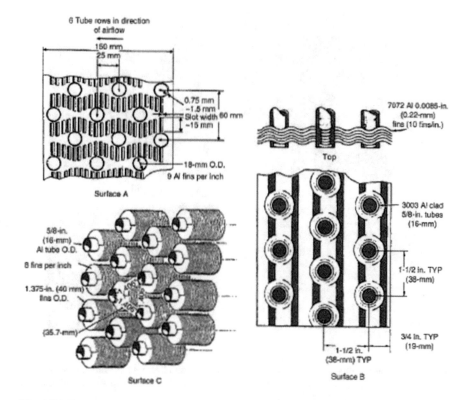

Fig. 4.25 Heat exchanger geometries tested by Kreid et al. (1983b)

fins provide natural drainage channels for a water film introduced at the top of the heat exchanger. Commercial coatings exist that reduce the contact angle of water on that surface and that promote surface wetting.

Controlling resistance in liquid phase is very important. The absorption process and the condensation of mixtures are such problems. Miller and Perez-Blanco (1993) tested specially configured tubes for use in the LiBr absorption process. They used three spine-fin tubes, one spirally indented tube, one spirally fluted tube and one ribbed tube (Fig. 4.26). Figure 4.27 shows that, for enhanced tubes, they absorbed mass increases as the falling film mass flow rate increases. However, for the smooth tube with film flow rates above 0.025 kg/s, the absorbed mass tends to level off.

Yuan et al. (2004) obtained LiBr absorption heat transfer coefficients in a horizontal staggered bundle of enhanced tubes. Three tube geometries—smooth, hydrophilic coated and axially fluted—were investigated. The effect of normal octyl alcohol additive was also tested. Figure 4.28 shows the heat transfer coefficient against solution flow rate. The enhancement by Marangoni convection by the additive is more effective for the axially fluted tube. Hoffmann et al. (1996) reported 20–40% increase of the heat transfer coefficient for a knurled tube over the smooth

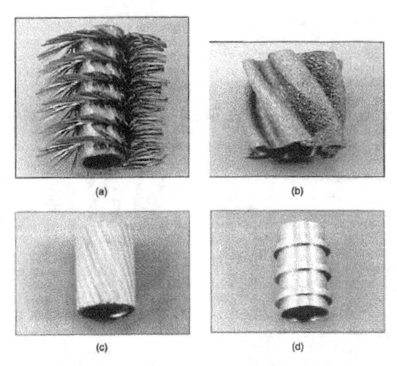

Fig. 4.26 Enhanced tubes tested by Miller and Perez-Blanco (1993) for the absorption of water vapour into LiBr solution: (**a**) spine fin tube, (**b**) spirally indented tube, (**c**) spirally fluted tube, (**d**) ribbed tube. From Miller and Perez-Blanco (1993)

tube for LiBr absorption on a horizontal tube bundle. Additives for example 1-octanol and 2-ethyl-1-hexanol increased the smooth tube bundle heat transfer coefficient by 60–140%. The increase was less (55–85%) for the knurled tube. Ishika et al. (1991) reported 30–80% heat transfer enhancement for the constant curvature surface (CCS) tubes.

An active enhancement test using vibrating screen was conducted by Tsuda and Perez-Blanco (2001) for steam absorption on vertical falling film of LiBr-water solution.

Pang et al. (2015) presented a detailed review on the simultaneous heat and mass transfer phenomenon in nanofluids. The thermal conductivity and heat transfer in nanofluids has been reviewed by Ozeriç et al. (2010), Lee et al. (2010), Yu et al. (2008), Wang et al. (2007), Chandrasekar and Suresh (2009), Sergis and Hardalupas (2011), Trisaksri and Wongwises (2007) and Godson et al. (2010).

With noncondensable gas in the condensing vapour, transfer impedance would exist for both phases, for both boiling and condensation of mixtures. Distillation columns involve two-phase flow and separation of mixtures in either tray towers or packed columns. Packed towers have a tortuous flow path for the gas phase and causes gas phase mixing. Thin liquid films exist on the packing.

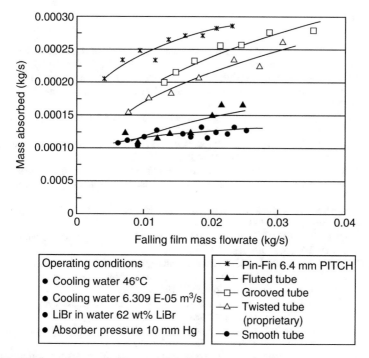

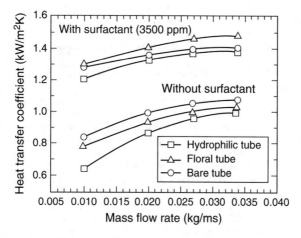

Fig. 4.27 Water vapour mass absorbed vs. falling film flow rate of LiBr solution for the enhanced tubes of Fig. 16.6. From Miller and Perez-Blanco (1993)

Fig. 4.28 Heat transfer coefficient vs. LiBr solution flow rate for enhanced tubes tested by Yoon et al. (2002). From Yoon et al. (2002)

References

Abid MF, Abdullah AN, Ahmad KM (2012) An experimental study of simultaneous heat and mass transfer in falling film and bubble mode absorbers. Petrol Sci Technol 30(5):467–477

Azizi Y, Bennacer R, Benhamou B, Galanis N, El-Ganaoui M (2007) Buoyancy effects on upward and downward laminar mixed convection heat and mass transfer in a vertical channel. Int J Numer Meth Heat Fluid Flow 17(3):333–353

Bassuoni M (2011) An experimental study of structured packing dehumidifier/regenerator operating with liquid desiccant. Energy 36(5):2628–2638

Bentley JM, Snyder TK, Glicksman LR, Rohsenow WM (1978) An experimental study of a unique wet/dry surface for cooling towers. J Heat Transf 100:520–526

Boukadida N, Nasrallah SB (2001) Mass and heat transfer during water evaporation in laminar flow inside a rectangular channel—validity of heat and mass transfer analogy. Int J Therm Sci 40 (1):67–81

Chandrasekar M, Suresh S (2009) A review on the mechanisms of heat transport in nanofluids. Heat Transf Eng 30(14):1136–1150

Chow LC, Chung JN (1983) Evaporation of water into a laminar stream of air and superheated steam. Int J Heat Mass Transf 26(3):373–380

Chung TW, Ghosh TK, Hines AL (1996) Comparison between random and structured packings for dehumidification of air by lithium chloride solutions in a packed column and their heat and mass transfer correlations. Ind Eng Chem Res 35(1):192–198

Duryodhan VS, Singh A, Singh SG, Agrawal A (2015) Convective heat transfer in diverging and converging micro-channels. Int J Heat Mass Transf 80:424–438

Fumo N, Goswami D (2002) Study of an aqueous lithium chloride desiccant system: air dehumidification and desiccant regeneration. Sol Energy 72(4):351–361

Gao W, Liu J, Cheng Y, Zhang X (2012) Experimental investigation on the heat and mass transfer between air and liquid desiccant in a cross-flow dehumidifier. Renew Energy 37(1):117–123

Godson L, Raja B, Lal DM, Wongwises S (2010) Enhancement of heat transfer using nanofluids e an overview. Renew Sust Energ Rev 14(2):629–641

Grossman G (1983) Simultaneous heat and mass transfer in film absorption under laminar flow. Int J Heat Mass Transf 26(3):357–371

Gu LD, Min JC, Tang YC (2018) Effects of mass transfer on heat and mass transfer characteristics between water surface and airstream. Int J Heat Mass Transf 122:1093–1102

Hoffmann L, Greiter I, Wagner A (1996) Experimental investigation of heat transfer in a horizontal tube falling film absorber with aqueous solutions of LiBr with and without surfactants. Int J Refrig 19(5):331–341

Huang K, Deng X (2018) Enhanced heat mass Transf of falling liquid films in vertical tubes. J Enhanc Heat Transf 25(1):79

Huang WJ, Deng XH, Zhou SH (2006) Mechanism of heat transfer enhancement for converging-diverging tube. J Fluid Machinery 2020

Hudina M, Sommer A (1988) Heat transfer and pressure drop measurements on tube and fin heat exchangers. In: Shah RK, Ganie EN, Yang KT (eds) Proc. 1st World Conf. on experimental heat transfer. Fluid mechanics and thermodynamics. Elsevier Science, New York, pp 1393–1400

Isshiki N, Ogawa K, Sasaki N, Funaro Y (1991) R & D of constant curvature surface (CCS) tubes for absorption heat exchangers. Proc Absorp Heat Pump Conf, Tokyo, Japan, pp 377–382

Iskra CR, Simonson CJ (2007) Convective mass transfer coefficient for a hydrodynamically developed airflow in a short rectangular duct. Int J Heat Mass Transf 50(11–12):2376–2393

Jang JH, Yan WM, huang CC (2005) Mixed convection heat transfer enhancement through film evaporation in inclined square ducts. Int J Heat Mass Transf 48(11):2117–2125

Johnson BM, Bartz JA, Alleman RT, Fricke HD, Price RE, Mcllroy K (1981) Development of an advanced concept of dry/wet cooling for power plants. Battelle Pacific Northwest Laboratories Report BN-SA-1296. Also presented at the American Power Conf April 27-29, Chicago, IL

Jung H, Hwang J, Jeon C (2014) An experimental study on performance improvement for an air source heat pump by alternate defrosting of outdoor heat exchanger. Int J Air-Conditioning Refrig 22(03):1450017

Kim JK, Park CW, Kang YT (2003) The effect of micro-scale surface treatment on heat and mass transfer performance for a falling film $H_2O/LiBr$ absorber. Int J Refrig 26(5):575–585

Koronaki IP, Christodoulaki RI, Papaefthimiou VD, Rogdakis ED (2013) Thermodynamic analysis of a counter flow adiabatic dehumidifier with different liquid desiccant materials. Appl Therm Eng 50(1):361–373

Kreid DK, Johnson BM, Faletti DW (1978) Approximate analysis of heat transfer from the surface of a wet finned heat exchanger. ASME paper 78-HT-26, ASME, New York.

Kreid DK, Hauser SG, Johnson BM (1983a) Investigation of combined heat and mass transfer from a wet heat exchanger. Part I: Analytical formulation. Proc ASME-JSME Joint Thermal Eng Conf. I:517–524

Kreid DK, Hauser SG, Johnson BM (1983b) Investigation of combined heat and mass transfer from a wet heat exchanger. Part 2: Experimental results. Proc ASME-JSME Joint Thermal Eng Conf 1:525–534

Kulankara S, Herold KE (2002) Surface tension of aqueous lithium bromide with heat/mass transfer enhancement additives: the effect of additive vapour transport. Int J Refrig 25(3):383–389

Kumar M, Ojha CSP, Saini JS (2014) Investigation of evaporative mass transfer with turbulent-forced convection air flow over roughness elements. J Hydrol Eng 19(11):06014004

Lee JH, Lee SH, Choi CJ, Jang SP, Choi SUS (2010) A review of thermal conductivity data, mechanisms and models for nanofluids. Int J Micro-nano Scale Transp 1(4):269–322

Lee JH, Jung CW, Chang YS, Chung JT, Kang YT (2016a) Nu and Sh correlations for LiCl solution and moist air in plate type dehumidifier. Int J Heat Mass Transf 100:433–444

Lee JH, Ro GH, Kang YT, Chang YS, Kim SC, Kim YL (2016b) Combined heat and mass transfer analysis for LiCl dehumidification process in a plate type heat exchanger. Appl Therm Eng 96:250–257

Li W, Wu XY, Luo Z, Webb RL (2011) Falling water film evaporation on newly-designed enhanced tube bundles. Int J Heat Mass Transf 54(13–14):2990–2997

Liu X, Jiang Y, Qu K (2007) Heat and mass transfer model of cross flow liquid desiccant air dehumidifier/regenerator. Energy Convers Manag 48(2):546–554

Luan SD, Liu CH, Shen ZQ (1993) An experimental investigation on falling film gas carrying evaporation in vertical tube. J Chem Eng Chin Univ 7(3):221–226

Ma X, Zhou XD, Lan Z, Song TY, Ji J (2007) Experimental investigation of enhancement of dropwise condensation heat transfer of steam-air mixture: falling droplet effect. J Enhan Heat Transf 14(4):295–305

Meyer T, Ziegler F (2014) Analytical solution for combined heat and mass transfer in laminar falling film absorption using first type boundary conditions at the interface. Int J Heat Mass Transf 73:141–151

Miller WA, Perez-Blanco H (1993) Vertical tube aqueous LiBr falling film absorption using advanced surfaces. Int Absorp Heat Pump Conf AES 31:185–202

Mittermaier M, Schulze P, Ziegler F (2014) A numerical model for combined heat and mass transfer in a laminar liquid falling film with simplified hydrodynamics. Int J Heat Mass Transf 70:990–1002

Moon C, Bansal P, Jain S (2009) New mass transfer performance data of a cross-flow liquid desiccant dehumidification system. Int J Refrig 32(3):524–533

Ozerinç S, Kakaç S, Yazıcıoglu AG (2010) Enhanced thermal conductivity of nanofluids: a state of the art review. Microfluid Nanofluid 8(2):145–170

Pang C, Lee JW, Kang YT (2015) Review on combined heat and mass transfer characteristics in nanofluids. Int J Therm Sci 87:49–67

Park CW, Kim SS, Cho HC et al (2003) Experimental correlation of falling film absorption heat transfer on micro-scale hatched tubes. Int J Refrig 26(7):758–763

Pehlivan H (2013) Experimental investigation of convection heat transfer in converging–diverging wall channels. Int J Heat Mass Transf 66:128–138

Poós T, Varju E (2017) Dimensionless evaporation rate from free water surface at tubular artificial flow. Energy Procedia 112:366–373

Rahamah A, Elsayed M, Al-Najem N (1998) A numerical solution for cooling and dehumidification of air by a falling desiccant film in parallel flow. Renew Energy 13(3):305–322

Raimundo AM, Gaspar AR, Oliveira AVM, Quintela DA (2014) Wind tunnel measurements and numerical simulations of water evaporation in forced convection airflow. Int J Therm Sci 86:28–40

Schwartze JP, Brocker S (2000) The evaporation of water into air of different humidities and the inversion temperature phenomenon. Int J Heat Mass Transf 43(10):1791–1800

Senshu T, Hatada T, Ishibane K (1981) Heat mass transfer performance of air coolers under wet conditions. ASHRAE Trans 87(Part 2):109–115

Sergis A, Hardalupas Y (2011) Anomalous heat transfer modes of nanofluids: a review based on statistical analysis. Nanoscale Res Lett 6(1):391–427

Sommer A (1984) Wasserverteilung aufuden Larnellen bespruhter FORGO-GLATT-Wurmetauscher, Beobachtungen and einem Plexiglas-Aluminum Modell. Report EIR-TM-23-84-09, Wurenlingen

Stegou-Sagia A (1996) An experimental study and a computer simulation of heat and mass transfer for three-dimensional humidification processes. Int J Numer Methods Biomed Eng 12(11):719–729

Sun P, Lin Z (1993) Heat transfer of falling film in vertical tube with superimposed vapor flow. J Chem Eng (China) 5:12–15

Talukdar P, Iskra CR, Simonson CJ (2008) Combined heat and mass transfer for laminar flow of moist air in a 3D rectangular duct: CFD simulation and validation with experimental data. Int J Heat Mass Transf 51(11):3091–3102

Tang R, Etzion Y (2004) Comparative studies on the water evaporation rate from a wetted surface and that from a free water surface. Build Environ 39(1):77–86

Tian H, Liu ZY, Ma YT (2012) Experimental research on falling film evaporating characteristic outside the horizontal enhanced tube. J Eng Thermophys 33(11):1924–1928

Trisaksri V, Wongwises S (2007) Critical review of heat transfer characteristics of nanofluids. Renew Sust Energ Rev 11(3):512–523

Tsuda H, Perez-Blanco H (2001) An experimental study of a vibrating screen as means of absorption enhancement. Int J Heat Mass Transf. 44:4087–4094

Volchkov EP, Leontiev AI, Makarova SN (2007) Finding the inversion temperature for water evaporation into an air–steam mixture. Int J Heat Mass Transf 50(11):2101–2106

Wan Y, Ren C, Xing L, Yang Y (2017a) Analysis of heat and mass transfer characteristics in vertical plate channels with falling film evaporation under uniform heat flux/uniform wall temperature boundary conditions. Int J Heat Mass Transf 108:1279–1284

Wan Y, Ren C, Yang Y, Xing L (2017b) Study on average Nusselt and Sherwood numbers in vertical plate channels with falling water film evaporation. Int J Heat Mass Transf 110:783–788

Wang BX, Zhang JT, Peng XF (2000) Experimental study on the dryout heat flux of falling liquid film. Int J Heat Mass Trans 43:1897–1903

Wang C-C, Chiang C-S, Lu D-C (1997) Visual observation of two-phase flow pattern of R-22, R-134a, and R-407C in a 6.5-mm smooth tube. Exp Thermal Fluid Sci 15:395–405

Wang Y, Deng X, Li Z (2007) Compound heat transfer enhancement of converged-diverged tube supported by twisted-leaves. J Chem Ind Eng-Chn 58(9):2190

Webb RL, Kim NY (2005) Principles of enhanced heat transfer. Taylor and Francis, New York

Webb RL (1988) Performance evaluation criteria for enhanced surface geometries used in two-phase heat exchangers. In: Shah RK, Subbarao EC, Mashelkar RA (eds) Heat transfer equipment. Hemisphere Pub. Corp, Washington, DC, pp 697–706

Webb RL (1991) Advances in shell side boiling of refrigerants. J. Institute of Refrigeration 87:75–86

Webb RL, Perez-Blanco H (1986) Enhancement of combined heat and mass transfer in a vertical tube heat and mass exchanger. J Heat Trans 108:70–75

Webb RL, Wanniarachchi AS, Rudy TM (1980) The effect of noncondensible gases on the performance of an R-11 centrifugal water chiller condenser. ASHRAE Trans 86(Part 2):170–184

Wei X, Duan B, Zhang X, Zhao Y, Yu M, Zheng Y (2017) Numerical simulation of heat and mass transfer in air-water direct contact using computational fluid dynamics. Procedia Eng 205 (Suppl C) 205:2537–2544

Wilke G, Herrmann G (1962) Ethylenebis (tri-phenyl-phosphine) nickel and analogous complexes. Angew Chem 74:693–694

Yang WJ, Clark DW (1975) Spray cooling of air-cooled compact heat exchangers. Int J Heat Mass Trans 18:311–317

Yang Z, Lin B, Zhang K, Lian Z (2015) Experimental study on mass transfer performances of the ultrasonic atomization liquid desiccant dehumidification system. Energ Buildings 93:126–136

Yin Y, Zheng B, Yang C, Zhang X (2015) A proposed compressed air drying method using pressurized liquid desiccant and experimental verification. Appl Energy 141:80–89

Yoshida T, Hyodo T (1970) Evaporation of water in air, humid air, and superheated steam. Ind Eng Chem Process Des Dev 9(2):207–214

Yoon JI, Kim E, Choi KH, Seol WS (2002) Heat transfer enhancement with a surfactant on horizontal bundle tubes of an absorber. Int J Heat Mass Trans 45:735–741

Yu W, France DM, Routbort JL, Choi SUS (2008) Review and comparison of nanofluid thermal conductivity and heat transfer enhancements. Heat Transf Eng 29(5):432–460

Yuan ZX, Yan XT, Ma CF (2004) A study of coupled convective heat and mass transfer from thin water film to moist air flow. Int Commun Heat Mass Transf 31(2):291–301

Zhang LZ (2011) An analytical solution to heat and mass transfer in hollow fiber membrane contactors for liquid desiccant air dehumidification. J Heat Transf 133(9):92001

Zhang L, Hihara E, Matsuoka F, Dang C (2010) Experimental analysis of mass transfer in adiabatic structured packing dehumidifier/regenerator with liquid desiccant. Int J Heat Mass Transf 53 (13):2856–2863

Zhang X, Wu J, Li Z (2019) Irreversibility characterization and analysis of coupled heat and mass transfer processes in an absorption system. Int J Heat Mass Transf 133:1121–1133

Zhao CY, Ji WT, Jin PH, Zhong YJ, Tao WQ (2018) Experimental study of the local and average falling film evaporation coefficients in a horizontal enhanced tube bundle using R134a. Appl Therm Eng 129:502–511

Chapter 5
Additives for Gases and Liquids

Liquid trace elements are added for boiling system, and additives for liquids include solid particles or gas bubbles in single-phase flows. For gases, additives are liquid droplets or solid particles; dilute phase will have gas-solid suspensions, and dense phase will have packed beds and fluidized beds.

5.1 Additives for Single-Phase Liquids

Solid particle additives may be water-chalk, water-coal, water-sand, water-aluminium, water-clay, water-copper, water-glass, water-graphite, ethylene glycol-graphite, kerosene-graphite (Kofanov 1964; Orr and Dallavalle 1954; Bonilla et al. 1953; Solamone and Newman 1955; Miller and Moulton 1956). Watkins et al. (1976) tested plastic beads in laminar oil flow. They observed a maximum enhancement of 40%. Fouling problem must be taken care of before using additive particles. Correlations for Kofanov (1964) are presented in Fig. 5.1.

Equations 5.1 and 5.2 give the correlations developed by Kofanov (1964)

$$Nu_d = 0.026\,Re_d^{0.8}\,\mathrm{Pr}^{0.4}\,F_p \tag{5.1}$$

$$F_p = \left(\frac{x_v}{1-x_v}\right)^{0.15}\left(\frac{\rho}{\rho_p}\right)^{0.15}\left(\frac{c_p}{c_{p.p}}\right)^{0.15}\left(\frac{d_i}{d_p}\right)^{0.02} \tag{5.2}$$

Ökten and Biyikoglu (2018a) examined the effects of air bubble injection on the heat transfer enhancement. The objective of Ökten and Biyikoglu (2018a) was to experimentally examine the heat transfer of air bubble injection systems within the thermal storage tanks. For this, they used 100-L plastic tanks with coiled tube in thermal tank including air injection pipe presented in Fig. 5.2. They placed

© The Author(s), under exclusive license to Springer Nature Switzerland AG 2020
S. K. Saha et al., *Electric Fields, Additives and Simultaneous Heat and Mass Transfer in Heat Transfer Enhancement*, SpringerBriefs in Applied Sciences and Technology, https://doi.org/10.1007/978-3-030-20773-1_5

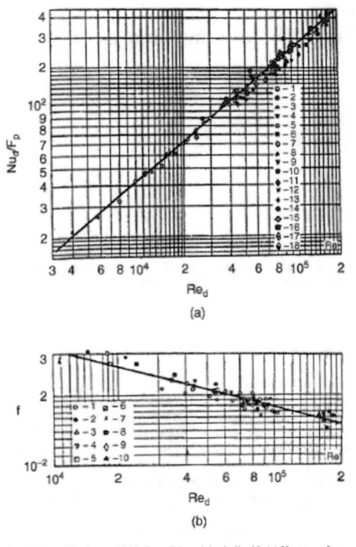

Fig. 5.1 Correlations of Kofanov (1964) for solid particles in liquid. (**a**) Heat transfer correlation, (**b**) friction factor correlation. See text for data sources (from Kofanov 1964)

thermocouples at eight positions, one at inlet and other at outlet of coiled tube. The experiments were performed for evaluating the overall heat transfer coefficient (K) between the storage fluid and the environment and between storage fluid and heat removal fluid. They carried out the experiments for both situations and presented their results in Fig. 5.3. This figure presents the heat transfer removal rate which was found to be seven times lesser with air bubble injection in comparison with the water tank without air bubble injection. Similarly, they analysed and presented energy equation solution:

Fig. 5.2 The placement of
the air injection pipe and
coiled tube in the tank
(Ökten and Biyikoglu
2018a, b)

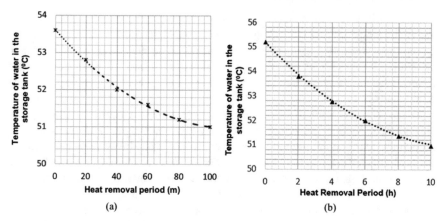

Fig. 5.3 Temperature variation of storage fluid during removal period (**a**) with air bubble injection
and (**b**) without air bubble injection (Ökten and Biyikoglu 2018a, b)

$$K^n = \frac{\left[\ln\left(\bar{T}_{sf}^n - T_o\right) - \ln\left(\bar{T}_{sf}^{n+1} - T_o\right) \right]}{A_t \Delta t / M_{sf} c} \tag{5.3}$$

They concluded that overall heat transfer coefficient (K) value was higher for the
tank equipped with air bubble injection system in comparison with the tank that did
not had such arrangement. They experimentally calculated the average values of K
for tank without air bubble injection system and tank having this system as
1207×10^{-3} and 5628×10^{-3} kW/m^2 °C, respectively. It is evident that four
times higher heat transfer coefficient was achieved by air bubble injection system.
Further, they calculated heat transfer between heat removal fluid and the storage
fluid and found that air bubble injection did not show any significant effect, and both
with and without air bubble injection systems were of comparable order. However,
they presented Fig. 5.4 which envelopes the amount of heat transfer of fluid at inlet

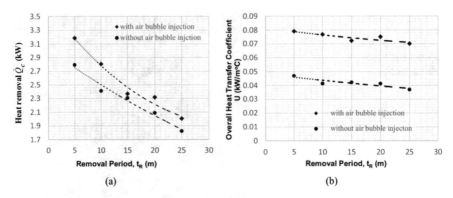

Fig. 5.4 The variation of (**a**) $\dot{Q}_C$ and (**b**) U with removal period with and without air bubble injection (Ökten and Biyikoglu 2018a, b)

and outlet temperatures. The influence of air bubble injection can be seen in plotted graph which confirms that the overall heat transfer coefficient for air bubble injection system is 1.7 times higher than that without this system, and this ratio is maintained as time goes on. They observed that air bubble injection worked significantly under minimum heat loss.

Dizaji (2014) investigated the application of air bubble injection for heat transfer enhancement with a horizontal double-pipe heat exchanger. The effect of two properties, e.g., vertical mobility of air bubbles and mainstream pressure of air for the heat transfer enhancement, was considered. Vertical mobility occurred due to buoyancy on bubble receding into the liquid, whereas airstream pressure provides extra motive power. They separately injected air bubble, so that no contact between air and water may occur before testing section. The objective of the study was to examine the effects of air bubble injection on the heat transfer parameters like heat transfer rate, number of heat transfer units (NTU) and effectiveness in a prescribed heat exchanger. Kwak and Oh (2000), Gabillet et al. (2002), Funfschilling and Li (2006), Ide et al. (2007), Kitagawa et al. (2008), Nouri and Sarreshtehdari (2009), Kitagawa et al. (2010), Saffari et al. (2013) and Samaroo et al. (2014) presented some recent investigations about air bubble injection method in various cross sections like annulus, tubes, plates, microchannels, etc. They concluded that, in laminar flow regime, fusion of two bubbles successively injected occurred. On the other hand, coalescence was not observed under turbulent regime. Also, bubble departure frequency was increased.

The whole testing setup has been presented in Fig. 5.5 which includes tubes, testing section, rotameters, water pump, condenser, compressor, evaporator, water tank and air pump. They listed the air bubble injection conditions in Table 5.1. They presented experimental correlation (5.4) to be valid in the range of $5000 \leq Re \leq 25{,}000$. They calculated the overall heat transfer coefficient with the typical correlation in eq. (5.5) and evaluated heat transfer coefficient by 'Wilson plots'. They concluded that air bubble injection effectively enhances heat transfer because of sticking of air bubbles to the inner part of inside tube and acts as insulator.

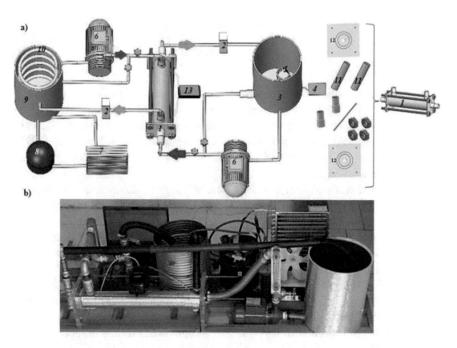

Fig. 5.5 (**a**) A schematic illustration of the test setup: 1-test section, 2-rotameter, 3-warm water tank, 4-dimmer and thermostat, 5-heater, 6- water pump, 7-condenser, 8-compressor, 9-cold water tank, 10-evaporator, 11-tube, 12-endplates of test section, 13-air pump and (**b**) experimental setup (Dizaji 2014)

Table 5.1 Experimental conditions and air bubble injection (Dizaji 2014)

Test no	n (mm)	d (mm)	Plastic tube situation	Range of inner tube, Re
1	Without air injection	–	–	5000–16,000
2	40	0.3	Inner tube	5000–16,000
3	40	0.7	Inner tube	5000–16,000
4	80	0.3	Inner tube	5000–16,000
5	80	0.7	Inner tube	5000–16,000
6	80	0.3	Outer tube	5000–16,000
7	80	0.7	Outer tube	5000–16,000

However, Nusselt number increased with increasing number of holes 'n' and with decreasing diameter of holes 'd' for inner tube air bubble injection. They observed that $n = 40$ with $d = 0.3$ provided better results than that of $n = 80$ and $d = 0.7$. So, the higher ratio of (n/d) was preferred for the enhancement in horizontal double-pipe heat exchanger. Also, they found that maximum effectiveness obtained was 40% with the air bubble injection. The increase in Reynolds number increased the NTU. It was observed that at higher Reynolds number or NTU, bigger bubbles ($d = 0.7$) had shown comparable or even better effectiveness than those with $d = 0.3$.

$$\text{Nu} = 1.84(\text{Re} - 1500)0.32 \, \text{Pr} \, 0.7 \tag{5.4}$$

$$U_{\exp} = \frac{Q_{ave}}{A\Delta T_{LMTD}} \tag{5.5}$$

Nano-sized metallic particles are used, and these increase the thermal conductivity of the flowing medium. To prevent agglomeration of nanoparticles, auxiliary activators and dispersants are used depending upon the properties of solutions and particles. Oleic acid laurate salt may be used as good dispersant. Choi (1995), Eastman et al. (1997), Xuan and Li (2000) and Li and Xuan (2000) give much useful information on nanofluids.

Very limited work on heat transfer characteristics of nanoparticle solutions had been reported till the last decade of the last century. However, since the early twenty-first century, there is a surge of focused research in nanoscience. Pak and Cho (1998) obtained heat transfer data of nanosize metallic oxide particles γ-Al_2O_3 and TiO_2, and the particle mean diameter was about 20 nm. The Nusselt number increase was due to the change of viscosity due to addition of particles.

Roy et al. (2006) used nanofluids for heat transfer augmentation from electronic component cooling equipment. The cooling system with radial flow was considered for numerical investigation. The single-phase flow assumption was made. They concluded that the addition of nanoparticles to base fluids results in increased wall shear stresses.

Early works on nanofluids were carried out by Boothroyd and Haque (1970, b), Kurosaki and Murasaki (1986), Murray (1994), Avila and Cervantes (1995, b), Sato et al. (1998), Wen and Ding (2004) and Ding and Wen (2005). The metallic Al_2O_3 nanoparticles were dispersed in water.

Heris et al. (2006) reported the performance of nanofluids for heat transfer enhancement in laminar flow through circular channel. The CuO-water nanofluid has been used for their study. Different percentage of volume fractions of CuO in water has been considered (0.2, 0.5, 1, 1.5, 2, 2.5 and 3). The variation of heat transfer coefficient with Peclet number has been presented in Fig. 5.6. Also, the results of heat transfer coefficients calculated from the correlation given by Sieder-Tate have also been plotted. It has been observed that the heat transfer coefficients obtained from the experiment were far higher than those predicted by the Sieder-Tate correlation. The increase in heat transfer enhancement ratio with Peclet number and nanoparticle concentration has been shown in Fig. 5.7. They observed an increase in enhancement ratio from 1.11 to 1.43 as the volume fraction of nanoparticles in the base fluid changed from 0.2 to 2.5%. They have also observed that at low Peclet numbers, there exists an optimum concentration at which heat transfer augmentation would be obtained. This is attributed to the increase in viscosity of the nanofluid with increase in nanoparticle concentration resulting in decreased heat transfer augmentation.

Li and Xuan (2000), Xuan and Li (2000, 2003), Xuan and Roetzel (2000), Khanafer et al. (2003), Wen and Ding (2004), Roy et al. (2004), Tsai et al. (2004) and Zeinali Heris et al. (2005) presented similar works.

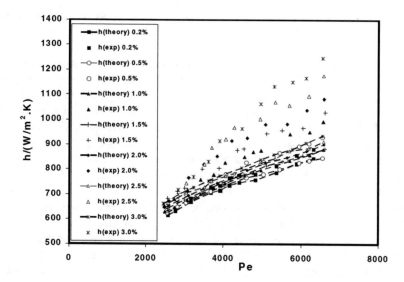

Fig. 5.6 Variation of heat transfer coefficient with Peclet number (Heris et al. 2006)

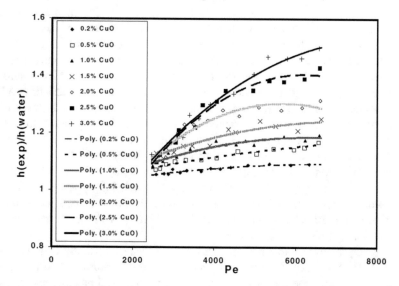

Fig. 5.7 Heat transfer enhancement ratio with Peclet number (Heris et al. 2006)

Akbarinia and Behzadmehr (2007, b) studied the effect of nanofluids in laminar mixed convection in horizontal curved tubes. Das et al. (2003a, b, c) presented the temperature dependence of thermal conductivity of nanofluids. Carbon nanotubes for heat transfer augmentation have been studied by Ding et al. (2006). Koo and Kleinstreuer (2005) reported the performance of nanofluids used in microheat sinks. Prasher et al. (2006) proposed a new convective-conductive model for thermal conductivity of nanofluids based on Brownian motion of nanoparticles. Zeinali Heris et al. (2007) used Al_2O_3-water nanofluid for convective heat transfer in circular cross-sectional tubes.

Esmaeili et al. (2010) studied the heat transfer performance of nanofluids in wavy channels. They used Al_2O_3-water nanofluids for augmentation of heat transfer. The finite volume technique was used to solve the uncoupled partial differential equations which govern the heat transfer and momentum transfer phenomena. The SIMPLE algorithm was used to solve the discretization equations. They concluded that the nanofluids resulted in enhanced heat transfer even for laminar flow regime. The thermal efficiency of nanoparticles was observed to increase with increase in Reynolds number and for larger channel curvatures.

Kowsary and Heyhat (2011) presented the nonhomogenous model for numerical investigation on augmentation of heat transfer using nanofluids. The top and bottom walls of the tube were symmetric isothermal heat sources. The γ-Al_2O_3-water nanofluids were used for the investigation of laminar flow heat transfer and wall shear stress characteristics. The Reynolds number range 600–2000 and the volume fraction of nanoparticle range 0.01–0.05 were considered. They concluded that the heat transfer enhancement in case of nonhomogeneous dispersion of nanoparticles was better than that in case of uniform dispersion. They observed that the particle concentration near the wall region was less due to the Brownian motion and thermophoresis effect which cause the skidding of nanoparticles. Another important observation was that the wall shear stress with nonhomogeneous dispersion was less than that with uniform dispersion. This can be attributed to the movement of nanoparticles from the wall towards the centre of the channel which results in increased temperature gradient and decreased viscosity of the flow near the wall.

The objective of the study Sandhu et al. (2018) was to experimentally examine the effects of nanoparticle concentration and its flow rate on augmentation of heat transfer. They considered alumina nanofluids with base fluid water (W) and water/ethylene glycol (W/EG) mixtures at ratios of 90:10, 80:20, 70:30, 60:40 and 50:50 for base fluid. Alumina nanoparticles sized 20 nm were additives. The alumina nanoparticles were spherical in shape which was concluded by scanning electron microscopy. They observed that maximum enhancement of thermal conductivity was obtained at 50 °C, and after that, it degraded. It may be due to nanoparticle agglomeration at higher temperature.

However, the overall thermal conductivity increased significantly with increased temperature. They experimented with nanofluids for measuring thermal performance. They varied the flow rates from 0.2 to 2 mL/min and nanoparticle concentration. The heat transfer performance with different base fluids is presented in Figs. 5.8 and 5.9. The increase in ethylene glycol ratio simultaneously increased the heat transfer rate. They observed that heat transfer enhancement for alumina-water was 6.20% which significantly increased to 36.5% for alumina with W/EG (50:50) base fluid having 0.1% nanoparticle volume fraction. The behaviour of Nusselt number against Reynolds number was much similar to that obtained with heat transfer coefficient with the flow rate. Nusselt number rises with increase in the concentration of the nanoparticles and Reynolds number. They presented the variation of friction factor related to alumina nanofluids against the increasing Reynolds

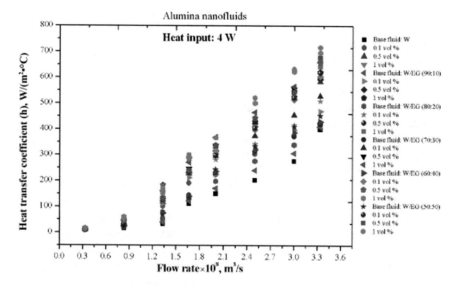

Fig. 5.8 Heat transfer coefficient of alumina nanofluids in microchannel (Sandhu et al. 2018)

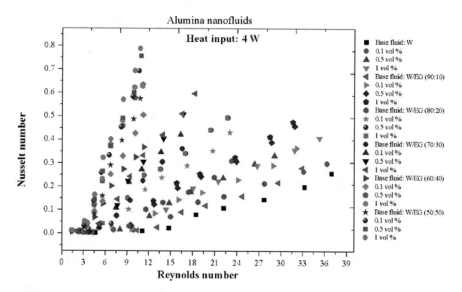

Fig. 5.9 Variation of the Nusselt number of the nanofluids with the Reynolds number (Sandhu et al. 2018)

number. They observed that pressure drop increased little with the nanofluid; it was maximum 5.7% for W/EG (50:50) base fluid with 1% volume concentration of alumina nanoparticles. Finally, they concluded that at 2 mL/min, the maximum enhancement of 36.5% was achieved due to 0.1% alumina nanoparticle volume concentration.

Sandhu et al. (2018) experimentally investigated the cooling performance of a microchannel heat sink with the influence of nanofluids. Similar investigation was performed by Chein and Chuang (2007). They investigated silicon-based microchannel heat sink performance using CuO-H_2O nanofluids. They found increased heat transfer coefficient with increase in nanoparticle concentrations. They calculated 4.2–5.7% and 10.2–15.8% increment in heat transfer with 0.15 and 0.26 volume per cent concentration, respectively.

Raisee and Moghaddami (2008) numerically investigated forced convection heat transfer characteristics for laminar flow through pipes having circular cross section. The γ-Al_2O_3 nanoparticles were dispersed in water. Maïga et al. (2004) and Koo and Kleinstreuer (2005) proposed simple model and the one which considers Brownian effect, respectively, for numerical investigation of heat transfer characteristics using nanofluids. They observed an increase in both wall heat transfer coefficient and wall shear stress with the addition of γ-Al_2O_3 nanoparticles to water. However, they concluded that the increase in heat transfer coefficient was higher compared to the increase in wall shear stress. They compared the two models proposed by Maïga et al. (2004) and Koo and Kleinstreuer (2005) and concluded that the reliability of the second model, which takes the effect of Brownian motion of nanoparticles in the base fluid, was greater. Also, they reported that the wall shear stress predicted by the first model was higher than that predicted by the second model. They observed minimum heat transfer enhancement of 10% for nanoparticle concentration of 1% and maximum of 30% enhancement for 4% nanoparticle concentration. Also, superior heat transfer enhancement using Al_2O_3-water nanofluids was reported at lower Reynolds number.

Chitra et al. (2015) investigated the heat transfer characteristics of $MgMnNiFe_2O_4$ dispersed in deionized water nanofluids for quenching in industrial application. Nanofluids are typically used for heat transfer enhancement in advance electronics. They used $MgMnNiFe_2O_4$/DIW-based nanofluids because of their enhanced thermal conductivity with the particle size ranging from 20 to 50 nm. They prepared nanofluids with various particle concentrations (1%, 2% and 4% by volume) by weighing and mixing. They presented Table 5.2 for the thermal conductivity enhancement which occurred due to the addition of different nanoparticles. The addition of low concentration of nanoparticles having high thermal conductivity explicitly increased the thermal conductivity of the fluid.

In this research work, they used KD2 Pro thermal property analyser with KS-1 sensor needle for the measurement of thermal conductivity. They measured conductivities for different nanoparticle concentrations (1%, 2% and 4%) and presented them in Table 5.3. They found that thermal conductivity of the $MgMnNiFe_2O_4$/DIW nanofluid decreased with the increment in the concentration of nickel. The experimental results for MgMnNi/DIW-based nanofluids provided higher thermal conductivity enhancements in comparison with these models. Figure 5.10 presents the experimental results obtained for thermal conductivity by varying the particle size and concentrations at different temperatures. They presented the volumetric specific heat for different concentrations in Table 5.4.

Table 5.2 Enhancement of thermal conductivity of water with the addition of nanoparticles, as reported in the literature (Chitra et al. 2015)

Particle material	Particle size (nm)	Concentration (vol. %)	Thermal conductivity ratio (K_{eff}/K_f)	Remarks	References
Cu	100	2.50–7.50	1.24–1.78	Laurate salt surfactant	Xuan and Li (2000)
	100–200	0.05	1.116	Spherical and square	Liu et al. (2004)
CuO	36	5	1.6	–	Eastman et al. (1997)
	23.6	1.00–3.41	1.03–1.12	–	Lee et al. (1974)
	23	4.50–9.70	1.18–1.36	–	Yu et al. (2008)
Al$_2$O$_3$	38.4	1.00–4.30	1.03–1.10	–	Lee et al. (1974)
	28	3.00–5.00	1.12–1.16	–	Yu et al. (2008)
	38.4	1.00–4.00	1.02–1.09	21 °C	Das et al. (2003a, b, c)
TiO$_2$	27	3.25–4.30	1.080–1.105	31.85 °C	Masuda et al. (1993)
	15	0.50–5.00	1.05–1.30	Sphere (CTAB)	Murshed et al. (2005)
Fe	10	0.2–0.55	1.14–1.18	–	Hong and Yang (2005)
MgMnNiFe$_2$O$_4$	0.50	1–4	0.99–1.12	20–90 °C	Present work

The variation of heat transfer coefficient with the Reynolds number is shown in Fig. 5.11. It was found that Nusselt number increased with increase in Reynolds number, and there was a strong and linear dependence on nanoparticle concentration. They provided better heat transfer performance in comparison with others; however, these nanofluids have a significant increase in thermal conductivity and decrease in viscosity with the variation in temperature. They found that the prescribed nanofluids possessed high thermal conductivity, leading to better energy efficiency and better performance with minimum operating cost. With the feature of better stability, these nanofluids (MgMnNiFe$_2$O$_4$/DIW, MgMnNi/DIW) restrict the rapid settling as well as minimize clogging on the walls of typical heat transfer devices. They proposed these fluids to be used in miniaturized system and vehicles as well as pumping equipment.

He et al. (2016) experimentally examined the effects of ZnO nanoparticle inclusions in ethylene glycol (EG) and deionized water mixture with different concentrations and studied the CHF of pool boiling heat transfer. They observed that enhancement in CHF was significant for all the nanofluids compared to the host fluid. Sarafraz et al. (2016) used synthesized zirconium oxide nanoparticles. They carried out the experiment using the nanofluids on a discoid copper heater for pool boiling CHF. They found 29% enhancement in boiling CHF. Cheedarala et al. (2016) used an eco-friendly nanofluid. This nanofluid was prepared by dispersion

Table 5.3 Thermal conductivity at different temperatures for various volume fractions of MgMnNi/DIW-based nanofluids (Chitra et al. 2015)

| S. No. | Temperature (°C) | Thermal conductivity ($Wm^{-1} K^{-1}$) $Mg_{0.40}Mn_{0.60}Ni_{0.00}$/DIW-based nanofluids | | |
		1%	2%	4%
1	35	0.610	0.612	0.615
2	50	0.612	0.614	0.616
3	65	0.613	0.616	0.618
4	80	0.615	0.617	0.621
S. No.	Temperature (°C)	Thermal conductivity ($Wm^{-1} K^{-1}$) $Mg_{0.40}Mn_{0.40}Ni_{0.20}$/DIW-based nanofluids		
		1%	2%	4%
1	35	0.639	0.642	0.644
2	50	0.639	0.644	0.645
3	65	0.641	0.644	0.648
4	80	0.642	0.645	0.652
S. No.	Temperature	Thermal conductivity ($Wm^{-1} K^{-1}$) $Mg_{0.40}Mn_{0.20}Ni_{0.40}$/DIW-based nanofluids		
		1%	2%	4%
1	35	0637	0.639	0.642
2	50	0.640	0.643	0.648
3	65	0.642	0.645	0.649
4	80	0.643	0.648	0.651

of cupric oxide (Cuo)-chitosan nanocomposite into deionized water. They concluded that there was 79% enhancement in CHF with 0.06% by weight of nanofluids. Ali et al. (2017) experimentally investigated the pool boiling heat transfer with the TiO_2-water nanofluid and achieved considerable enhancement in CHF. Ham et al. (2017) performed experimental investigation on pool boiling where combined effect of surface roughness and Al_2O_3-deionized water nanofluids was involved.

They claimed significant improvement of 224.8% and 138.5% with the surface roughness of 177.5 mm and 292.8 nm respectively. However, they observed CHF deterioration under high volume concentrations for both the surface roughness levels. Ciloglu (2017) investigated nucleate pool boiling heat transfer with SiO_2-based deionized water nanofluid. He reported 45% increase in CHF with the increased concentration and explained that it happens due to decrement in contact angle between the liquid droplets and the surface. Similarly, Choi et al. (2017) found 40% improvement in subcooled flow boiling CHF with the magnetic nanoparticles suspended in deionized water. Sulaiman et al. (2016) tested the heat transfer performance associated with the saturated pool boiling of nanofluids with TiO_2, Al_2O_3 and SiO_2 nanoparticles. They varied additive particle sizes from coarse to fine as well as varied their concentration. They achieved enhancement for all nanofluids over that for using only base fluid.

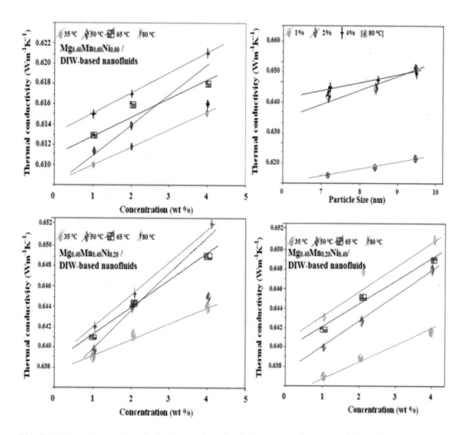

Fig. 5.10 Experimental results for thermal conductivity values of MgMnNi/DIW-based nanofluids at different particle sizes (top right); experimental results for thermal conductivity values of MgMnNi/DIW-based nanofluids at various concentrations with different temperatures (bottom) (Chitra et al. 2015)

Kamel et al. (2018) presented a review on the developments in boiling critical heat flux due to inclusion of nanofluids. They considered both pool boiling and convective flow boiling. Their review provided the significance of nanofluids in achieving high heat flux under small temperature gradient circumstances, which leads to a safer and more durable heat exchanger performance. So, high heat flux can be achieved easily for smaller wall superheat temperatures. The main reason involved in improvement of CHF with nanofluids is wettability of heating surface. It is caused by decreased contact angle and capillary wicking forces. They suggested that nanoparticle suspension in bulk fluid should be considered. They found little amount of literatures on CHF of convective flow boiling heat transfer. This was due to very complex setup requirements for such type of boiling. They suggested that future of CHF improvement involves use of nanofluid as it may provide excellent heat flux density. Application of magnetic field with ferro-fluids can be effective in the augmentation of heat transfer. Novel efficient eco-friendly nanocomposites

Table 5.4 Volumetric specific heat measurements for different concentrations of nanoparticles at different temperatures (Chitra et al. 2015)

Temperature (°C)	Volumetric specific heat (MJ/m³·K) at 1%	
	Experimental results	Calculated values
10	3.65	3.61
25	3.74	3.65
35	3.79	3.70
50	3.88	3.84
65	3.95	3.89
90	3.11	3.10
Temperature (°C)	Volumetric specific heat (MJ/m³·K) at 2%	
	Experimental results	Calculated values
10	3.63	5.59
25	3.72	5.62
35	3.77	5.68
50	3.87	5.83
65	3.94	5.88
90	3.11	5.10
Temperature (°C)	Volumetric specific heat (MJ/m³·K) at 4%	
	Experimental results	Calculated values
10	3.60	3.54
25	3.68	3.61
35	3.73	3.65
50	3.82	3.79
65	3.90	3.84
90	3.104	3.098

should be targeted. Hybrid nanofluids with two or more nanoparticles can be investigated; some of literatures were already presented for the enhancement for single phase.

A lot of efforts have been applied for examining the performance of boiling heat transfer with the nanofluids as it has a wide range of application. All these articles and literatures concluded that generation of porous layer due to nanofluid maximize the CHF. Some researchers You et al. 2003; Ham et al. 2017) claimed that they achieved 200% enhancement in CHF with low concentration of nanoparticles.

Allen and Cooper (1987) worked on the application of electric field on the boiling of R-114, and Ogata and Yabe (1993) investigated pool boiling of R-11 and mixture of R-11 with ethanol. Similarly, Ogata and Yabe (1991), Paper et al. (1993), Kwak and Oh (2000), Neve and Yan (1996) and Cho et al. (1996) presented the EHD heat transfer enhancement technique for boiling on horizontal surfaces. Kweon et al. (1998) experimentally applied non-uniform electric field and three different types of electrodes for the investigation of bubble behaviour. They observed that bubble departure volume reduced with AC electric field significantly rather than with DC electric field and it dropped suddenly at critical voltage. Further, Kweon et al. (1998) examined the influence of electric field on nucleate boiling with R-113 fluid. They observed that boiling parameter was significantly influenced under non-uniform

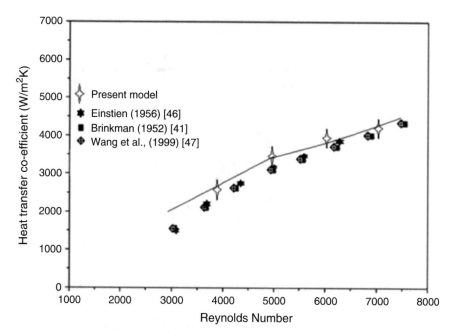

Fig. 5.11 Change in heat transfer coefficient with Reynolds number for MgMnNi/DIW-based nanofluid (Chitra et al. 2015)

electric field, and latent heat conjugated with bubbles almost corresponds to total heat at high voltage. Liu et al. (2004) discussed EHD enhancement technique with the electrode polarity effect on boiling heat transfer within a vertical tube where R-123 was used as working fluid. They concluded that positive high voltage yielded higher enhancement at the less amount of electric field strength than that of the negative high voltage.

Grassi and Testi (2006) presented experimental results related to the pool film boiling heat transfer with testing on wires at atmospheric pressure. They used R-113 ($\varepsilon_r = 2.41$) and Vertrel XF ($\varepsilon_r = 6.72$) for testing electric permittivity in this process. They obtained increased heat flux at a given wall superheat and presented it in Figs. 5.12 and 5.13. Between the two fluids, the Vertrel XF (polar fluid) performed better than weak polar R-113 liquid. They concluded that oscillation wavelength and wall superheat were strongly interconnected. The transition between regimes revealed that wall superheat overpowers the heat flux, and they detected the transition from one-dimensional to a two-dimensional regime.

Pal and Bhattacharyya (2018) numerically investigated the heat transfer enhancement of copper-water nanofluid within a heated pattern channel. They used surface-mounted protrusions on one wall in a horizontal tube for numerical analysis. Their objective was to increase the rate of heat transfer by additives and modulation of channel inner peripheral walls. They considered ribs in cosine wave form with flat recedes against each protrusion. The block or triangular protrusion caused high

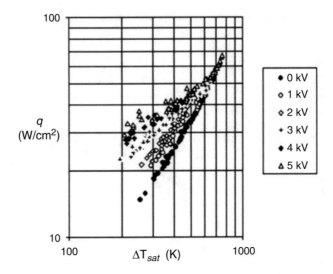

Fig. 5.12 Film boiling curve of R113 for different high-voltage values (Grassi and Testi 2006)

Fig. 5.13 Film boiling curve of Vertrel XF for different high-voltage values (Grassi and Testi 2006)

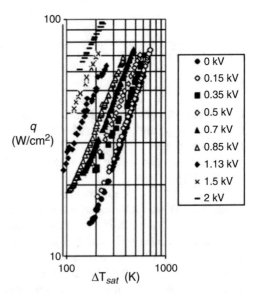

pressure drop. So, they considered blunt cosine wave form ribs. They placed ribs along transverse direction of flow and calculated the effect of nanoparticle volume fraction as well as rib. They considered parameters like entropy generation, friction factor and surface area to heat transfer enhancement ratio for evaluation. Similarly, Heidary and Kermani (2012), Akbari et al. (2016), Gravndyan et al. (2017), Arani et al. (2017) examined different aspects of nanofluid influence on heat transfer enhancement.

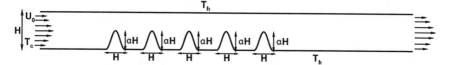

Fig. 5.14 Schematic diagram of physical space and coordinates (Pal and Bhattacharyya 2018)

Table 5.5 Thermo-physical properties of water and copper (Pal and Bhattacharyya 2018)

Parameter	Water	Copper
c_p (J/kgK)	4179	383
ρ (kg/m^3)	997.1	8954
k (W/mK)	0.6	400
β (K^{-1})	2.1×10^{-4}	1.67×10^{-5}

Pal and Bhattacharya (2018) presented their model in Fig. 5.14. They assumed that working fluid is Newtonian having constant properties and that the flow is incompressible, laminar, 2D steady state flow. Negligible radiative heat transfer and no chemical reaction were considered for numerical analysis. They presented thermo-physical properties of both water and copper in Table 5.5. They used finite volume method and divided computational domain into Cartesian cells. They validated their results with Wang and Chen (2002) and Heidary and Kermani (2012). They simulated for forced convection in a channel with Copper-water nanofluids and blunt rib where width was much bigger than its height. They varied concentration in the range of 0.0–0.4.

They observed that the Nusselt number (Nu$_{avg}$) and total entropy generation (S$_{tot}$) increased monotonically with the increased Reynolds number. Increased Reynolds number dictated thinner boundary layer on wall causing stronger temperature gradient and thus increased average Nusselt number. At the Reynolds number $Re = 50$, only volume fraction effectively worked, whereas at $Re = 200$, 16% enhancement was achieved due to the combined effect. They presented the influence of rib and nanoparticle volume fraction on the enhancement of heat transfer, entropy generation and friction factor in Fig. 5.15. They presented the results for Reynolds number ranging in between 60 and 250. They defined the ratio of average Nusselt number for water and smooth channel as $K = {Nu_{av}}/{Nu_{av}^P}$. Similarly, for entropy generation $\tau = {S_{tot}}/{S_{tot}^P}$, friction factor ratio was defined as ${f}/{f^P} = {\Delta p}/{\Delta p^P}$ where Δp^P is the corresponding pressure drop for clear fluid ($\Phi = 0$).

They presented Fig. 5.15 for enhancement factor, pressure drop ratio and entropy generation ratio against the Reynolds number for one rib, three ribs and five ribs in a single graph for ease of comparison in Fig. 5.15a, b and c, respectively. They included the volume fraction factor $\Phi = 0.0$ with solid lines and $\Phi = 0.04$ with dashed lines. Maximum enhancement was achieved at Reynolds number 100 because of recirculation zones in-between adjacent ribs. With increase in nanoparticle inclusions, enhancement factor increased. The net entropy generation (τ) increased with Reynolds number up to 150 and saturated for further increase in

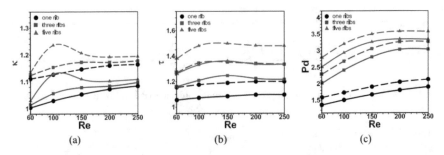

Fig. 5.15 Variation of (**a**) κ ($= Nu_{avg}/Nu_{av}^{P}$), (**b**) τ ($= S_{tot}/S_{tot}^{P}$) and (**c**) Pd ($= f/f^{P}$) with Re at different φ where solid and dashed lines are for $\varphi = 0.0$ and $\varphi = 0.04$ cases, respectively (Pal and Bhattacharyya 2018)

Reynolds number. The additives and increased number of ribs enhanced heat transfer to a great extent. They observed that friction factor increased linearly with Reynolds number in laminar regime, whereas non-linear variation occurred with multiple ribs.

Enhancement occurs when a gas bubbles through a stationary liquid such as in the case of nucleate boiling on a surface (Tamari and Nishikawa 1976; Kenning and Kao 1972). This enhancement is due to liquid agitation on the surface.

Suspension in dilute polymer and surfactant solutions reduce friction. Long fibres dampens fluid turbulence and reduce the turbulent kinetic energy dissipation (Lee et al. 1974). These fibre suspensions may be few hundred parts per million (ppm). Moyls and Sabersky (1975) investigated the effects of dilute suspensions of asbestos fibres (300 ppm) in an aqueous solution containing a polymer. Fossa and Tagliafico (1995) tested a dilute polymer-water solution in an annular passage with ribs machined on the outer surface of the inner tube.

Surfactants are also used to reduce the pressure drop, typically accompanied by the heat transfer reduction. Qi et al. (2001) tested two surfactant additives (2300 ppm Ethoqued T13-50 and 1500 ppm SPE98330) in a smooth tube and in a four-start spirally indented tube.

The effect of surfactants added to the alumina-water nanofluids on the enhancement of thermal conductivity was studied by Li et al. (2009). They used sodium dodecyl benzene sulfonate (SDBS) as the surfactant. SDBS is an anionic surfactant. They prepared the nanofluids by using a two-step method in which the first step was to prepare nanoparticles and in the second step the nanoparticles were dispersed in the base fluid. The preparation of nanofluids demands stable and durable suspension, low agglomeration of nanoparticles in the base fluid, etc. There are various techniques used for changing the surface properties of suspensions and for the prevention of cluster formation. These techniques include changing the pH value of suspensions, using surface activators and dispersants and using ultrasonic vibration.

The stability of nanofluids is expressed in terms of zeta potential. The zeta potential for the given nanofluids using SDBS dispersant has been shown in

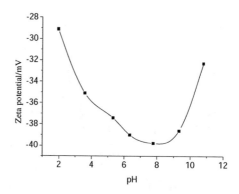

Fig. 5.16 Zeta potential for the given nanofluids using SDBS dispersant (Li et al. 2009

Fig. 5.16. The pH range of 8–9 has been reported to be the operating pH for nanofluids with SDBS dispersant. An increment in thermal conductivity of nanofluids over that of the base fluid at 0.0015 weight concentration of nanoparticles was observed by using Hamilton–Crosser model. The thermal conductivity and electrical conductivity measurements for nanofluids were presented by Babcsan et al. (2003). Cheol et al. (2002) described the dispersion of carbon nanotubes using sonication. The dependence of thermal conductivity on temperature was reported by Das et al. (2003a, b, c). Eastman et al. (1997, 2001), Elimelech et al. (1995), Hamilton and Crosser (1962), Hong et al. (2006), Rodríguéz-Perez et al. (2003) and Zhu et al. (2007) used nanofluids for heat transfer enhancement.

Liao et al. (2010) studied flow drag and heat transfer characteristics of CuO nanoparticles suspended in base fluids flowing in a small tube. They also studied the performance of nanofluids with surfactants. The effect of fluid temperature and addition of surfactants on flow drag and heat transfer characteristics have been presented in detail. The nanoparticles added to deionized water served as one of the two working fluids considered for their study. In order to obtain a homogenous mixture, the nanofluids were placed in an ultrasonic water bath and were subjected to oscillations for about 10 h.

This working fluid was referred to as nanoparticle suspension throughout their study. The sodium dodecyl benzene sulfonate (SDBS) surfactant was added to the deionized water along with the nanoparticles. This is followed by the procedure of obtaining homogenous mixture, in the above-stated manner. This working fluid was referred to as nanofluids in order to differentiate it from the other working fluid type (suspensions). The surfactant weight concentration was 1% for the entire study. The details of the working fluids (nanoparticle suspensions and nanofluids) have been tabulated in Table 5.6.

The flow drag in terms of pressure drop versus flow velocity and friction factor versus Reynolds number has been shown in Fig. 5.17a, b, respectively. They observed an increase in pressure drop for suspensions compared to that of water for laminar flow. Also, the increase in pressure drop with the concentration of

Table 5.6 Details of the working fluids (nanoparticle suspensions and nanofluids) (Liao et al. 2010)

Temperature	$25 \pm 2^{\circ}C$	$58 \pm 2^{\circ}C$		
Working fluid				
Water	In a new tube	In a new tube		
	In an old tube			
CuO-water nanosuspension	Concentration of CuO	Concentration of SDBS	Concentration of CuO	Concentration of SDBS
	1 wt.%	0	2 wt.%	0
	2 wt.%	0		
	4 wt.%	0		
CuO-water nanofluid	2 wt.%	1 wt.%		

nanoparticles was observed. It was also reported that the pressure drop for nanofluids has increased with the addition of SDBS surfactant. This increase can be attributed to the increase in viscosity due to the addition of nanoparticles and surfactants. On the other hand, the pressure drop in turbulent flow was observed to be lower in the case of suspensions compared to the case of water. Also, the pressure drop in old tube was lower than that for new tube, explaining the effect of sedimentation on fluid flow drag.

The dimensionless heat transfer for suspensions and nanofluids in laminar and turbulent flow regions has been presented in Fig. 5.18. They concluded that the surfactants do not affect the heat transfer. They reported that the enhancement in heat transfer at higher fluid temperature was much greater than that observed at lower fluid temperature. They also observed that the integrated enhancing ratio for suspensions at 25 °C was in turbulent flow while it was less than 1 for laminar flow. In the case of nanofluids, the integrated enhancing ratio was less than 1 for both laminar and turbulent flow regimes. The integrated enhancing ratio corresponding to the suspensions at 58 °C was 1.4 and 1.1 in turbulent flow and laminar flow, respectively.

The heat transfer augmentation using nanofluids for turbulent flow was studied by Zhang et al. (2010). The CuO nanoparticles of different sizes (23, 51 and 76 nm) were dispersed in water, and forced convection heat transfer under constant heat flux boundary condition was investigated. They also used SDBS dispersant in order to obtain uniform dispersion. They chose hydrochloric acid and sodium hydroxide in order to vary the pH value of the dispersed fluid. The zeta potential was reported to be maximum corresponding to the pH value 8 and small particle size resulting in most stable nanofluids. They have also developed a novel correlation for the effective thermal conductivity of the nanofluids. The correlation was developed by considering the cluster distribution and surface adsorption of the nanoparticles. They reported increase in Nusselt number with nanoparticle size for high Reynolds number. They concluded that the turbulent heat transfer was due to the irregular pulse of combined convection.

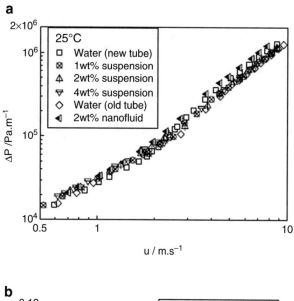

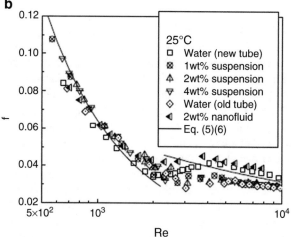

Fig. 5.17 Flow drag characteristics of the tested liquids: (**a**) relation between the pressure drop and flow velocity; (**b**) relation between the flow drag and Re number (Liao et al. 2010)

Sato et al. (1998) and Kumada et al. (2002) investigated the effect of turbulence promoters, for example fencer, sawtooth plates, porous plates, and vortex generators, in channel flow of an aqueous solution containing a surfactant. Li et al. (2001) installed fine mesh structures at the frontal part of the interested area, and this novel method was used to increase heat transfer by breaking the rod-like microcellular structure of the surfactant. The mesh includes high shear stress, and the turbulence intensity in the near-wall region is reduced.

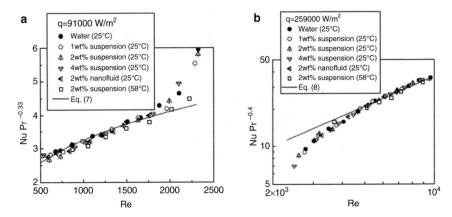

Fig. 5.18 Dimensionless heat transfer for suspensions and nanofluids in laminar and turbulent flow regions (Liao et al. 2010)

5.2 Additives for Single-Phase Gases

Flow of gas-solid suspension inducts and fluidized bed require solid additives. Gabor and Botterill (1985) reviewed the literature on fluidized beds. Figure 5.19 deals with the applicable experimental information on gas-solid suspensions flowing inside tubes. The correlation in Fig. 5.20 is that due to Sadek (1972). Gabor and Botterill (1985) gave heat transfer correlations for horizontal bundles of plain and finned tubes. Several studies have been performed to measure the advantages offered by finned and rough surfaces in fluidized beds. Use of an enhanced surface is considered to be compound enhancement, because two enhancement techniques are combined. Petrie et al. (1968) worked with finned tubes in a horizontal fluidized bed and examined the effect of fin pitch for plain tubes. They measured a gas-side enhancement ratio. Bartel and Genetti (1973) measured heat transfer coefficients for horizontal bundles of steel tubes having radial, plain aluminium fins. Krause and Peters (1983) tested steel segmented finned tubes in a fluidized bed (Table 5.7). Three fin heights were tested. Experimental results had been presented as ratios, relative to the plain tube bundle. The highest enhancement ratio of 5.82 was obtained with the largest particle size and the highest fin. The lowest fins, however, gave the best performance for the smaller particle size. Chen and Withers (1978) tested vertical integral fin tubes inside a vertical tube using glass particles, and Table 5.8 shows their results. Grenwal and Saxena (1979) used tubes with closely spaced v-thread or knurled roughness in a horizontal fluidized bed. The roughness height was 1.07 mm or less. The best roughness provided 40% higher heat transfer coefficient than did the plain tube.

Addition of water droplets to the airstream is the classic example of liquid additive; Thomas and Sunderland (1970), Bhatti and Savery (1975), Kosky (1976) and Nishikawa and Takase (1979) are the relevant literature on liquid additives. The

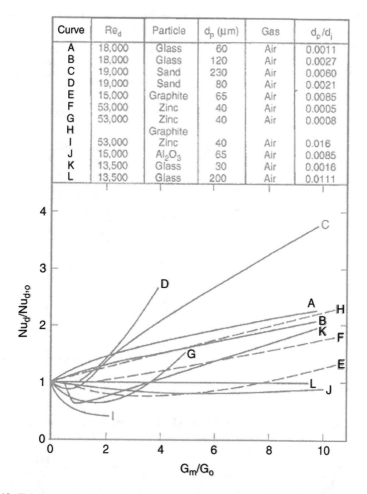

Fig. 5.19 Enhancement ratio vs. G_m/G_0 loading ratio for various solid particle mixtures

water wets the heat transfer surface, providing evaporation from the water film surface into the airstream. Very high enhancement can be provided, if the surface can be fully wetted.

5.3   Additives for Boiling, Condensation and Absorption

Lowery and Westwater (1959), Gannett and Williams (1971), Wasekar and Manglik (1999, 2000), Tzan and Yang (1990), Ammerman and You (1996), Yang et al. (2002), Wu et al. (1998), Wen and Wang (2002), Kotchaphakdee and Williams (1970), Paul and Abdel-Khalik (1983), Wang and Hartnett (1992), Wu and Yang

Fig. 5.20 Correlation of gas-solid suspension heat transfer data for turbulent flow in tubes (from Furchi et al. 1988)

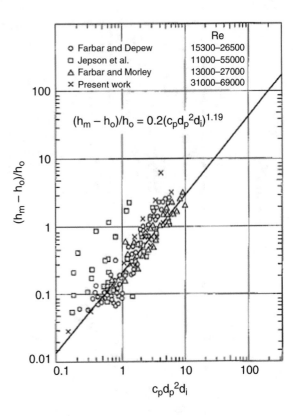

Table 5.7 Segmented finned tube performance in horizontal fluidized bed bundle $(G/G_{11''} = 1.2)$

	$d_p = 0.21$ mm	$d_p = 0.43$ mm				
e_o (mm)	η_f	h/h_s	hA/h_sA_s	η_f	h/h_s	hA/h_sA_s
4.76	0.78	0.88	3.74	0.81	0.89	3.80
8.33	0.65	0.72	4.93	0.64	0.81	5.57
11.11	0.66	0.34	3.03	0.68	0.65	5.82

Table 5.8 Performance of vertical integral fin tubes in fluidized bed (glass particles, $dP = 0.25$ mm)

E_{ho}	Fins/m	e_o (mm)	A/A_s	$\eta h/h_s$
0.60	197	3.18	2.22	0.70
1.50	354	1.57	1.88	1.00
2.30	433	3.18	3.87	0.90
1.80	748	3.06	3.06	0.80

(1992), van Wijk et al. (1956), van Stralen (1959) and Bergles and Scarola (1966) are some important literatures on additives used in boiling. The only known promoters that are indefinitely durable are certain coatings on the base surface. These are gold and Teflon. Figures 5.21 and 5.22 show some useful results. The objective for condensation and adsorption is to promote dropwise condensation. Griffith

Fig. 5.21 Pool boiling curves of aqueous solution of SLS on a 3.35-mm stainless heater (Webb and Kim 2005)

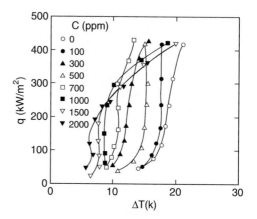

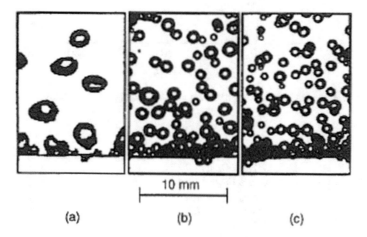

Fig. 5.22 Flow visualization photographs illustrating the effect of surfactant on the bubble generation from 0.39-mm platinum wire in SDS aqueous solution: (**a**) 0 ppm, (**b**) 500 ppm, (**c**) 1000 ppm

(1985), Tanasawa (1978), Kim et al. (1996, 2001), Depew and Reisbig (1964), Rush (1968), Rush et al. (1991) and Ziegler and Grossman (1996) may be referred for more information on additives for condensation and absorption.

Wen et al. (2006) investigated the performance of TiO_2 nanoparticles suspended in water for heat transfer enhancement in pool boiling. Masuda et al. (1993, 1993) and Choi (1995) were the pioneers in the field of nanofluids. The term nanofluids were coined by Choi (1995). The variation of heat transfer coefficient with heat flux has been shown in Fig. 5.23, for both base fluid and nanofluids. Also, the enhancement ratio variation with heat flux has been presented in Fig. 5.24. They observed 50% increase in heat transfer over that of base fluid for nanofluid with 0.72% nanoparticles by volume. The results obtained have also agreed well with the results of Rohsenow (1952).

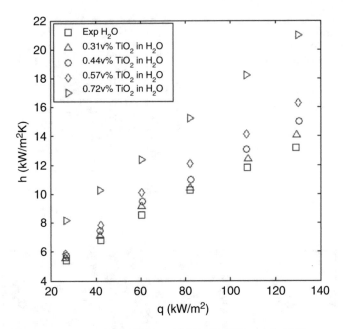

Fig. 5.23 Variation of heat transfer coefficient with heat flux (Wen et al. 2006)

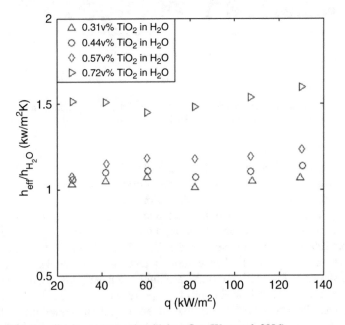

Fig. 5.24 Variation of enhancement ratio with heat flux (Wen et al. 2006)

The nucleate pool boiling heat transfer phenomenon using nanofluid was studied by Das et al. (2003a, b, c). They studied heat transfer on a cylindrical cartridge heater surface. You et al. (2003), Bang and Chang (2004), Tu et al. (2004) and Witharana (2003) carried out heat transfer enhancement investigations using different nanofluids.

Lv and Liu (2008) experimentally studied the effect of nanoparticle suspension on boiling heat transfer augmentation in vertical tubes immersed in saturated nanofluids. The base fluid was considered to be water. The effect of nanoparticle concentration on boiling heat transfer augmentation and critical heat flux has been investigated. They observed that the nucleate boiling heat transfer in small tubes is quite complex than conventional nucleate pool boiling. This is because the vapour bubble cannot freely rise from the wall as they are restricted in a narrow space. The bubbles are thus crowded and displace more liquid. In this case, the heat transfer is augmented due to the formation of thin liquid film at the interface of tube wall and the coalesced bubbles. Lv and Liu (2008) observed that the boiling heat transfer for tubes immersed in nanofluids was superior to that for the tubes immersed in water. The increase was about 50–80% at 1% mass concentration of nanoparticles.

Also, about 40% increase in critical heat flux was observed with nanofluids over that with water. This increase was observed for lower concentrations of nanoparticles. The increase in critical heat flux with nanoparticle concentration was observed up to 1% concentration of nanoparticles after which constant heat flux values are reported. They have also developed empirical correlations for predicting the critical heat flux. Katto and Kawamura (1981), Monde and Yamaji (1990) and Liu (2001) have also worked on boiling heat transfer enhancement in vertical tubes submerged in saturated liquids.

The application of nanofluids for heat transfer in an evaporator in a capillary pumped loop (CPL) was studied by Lv and Liu (2010). They investigated the effect of nanoparticle parameters such as mass concentration, type and size on heat transfer coefficient and maximum heat flux in the evaporator in CPL. They observed an increase in heat transfer coefficient and maximum heat flux due to the addition of nanoparticles and reported that optimal parameters of nanoparticles result in maximum heat transfer augmentation. They studied the performance of Cu and CuO nanoparticles and observed that 1% and 0.5% mass concentrations were the optimal values for Cu and CuO nanoparticles, respectively. They have concluded that the enhancement in heat flux ratio were maximum 16%, 14.5% and 12% corresponding to the 20 nm Cu particles, 50 nm Cu particles and 50 nm CuO particles, respectively.

Hestroni et al. (2006) reported the effect of surfactants in heat transfer characteristics of boiling phenomenon. The boiling heat transfer augmentation and the bubble dynamics for various concentrations of the surfactant have been studied. The high-speed video recording technique has been used in order to study the bubble nucleation, bubble growth and bubble departure. They observed that the bubble growth in surfactant solutions is effected by different parameters like interfacial properties, chemistry and nature of the additives, foaming, etc. They used high-speed camera and recorded the bubble growth at a speed of 1000 frames per second. From the figure, they observed the bubble growth in both water and alkyl surfactant solution. In case of water, regular bubble shape is observed. During the bubble growth, the shape is altered in order to avoid necking. Once the necking starts, the centre of the bubble clusters is seen where the bubbles are found to be adjacent to one another. Thus, bubble coalescence is observed. The coalescence between vapour particles reduces with surfactant additives. They concluded that in surfactant solution, the bubbles were smaller and covered the surface at a faster rate compared to those in the case of water. Increase in heat transfer rates with increase in the concentration of the surfactants was reported. The solutions with different surfactant concentrations have same boiling curve at different subcooling levels.

Wasekar and Manglik (2017) investigated the enhancement of nucleate boiling heat transfer with the addition of surfactant and polymer in water at low concentrations. They found the heat transfer enhancement was influenced by additive concentration, its type and chemistry, wall heat flux and heater geometry. Yoo (1974), Cho and Hartnett (1982), Irvine and Kami (1987) reported that additives did not significantly affect the thermo-physical properties of the solvent except surface tension and apparent viscosity. Table 5.9 shows the literature on heat transfer enhancement classified according to the types of additives used, heating geometry, heat flux and their effect on heat transfer. They observed that surface tension and apparent viscosity were primarily influenced by small amounts of surfactant or polymer additives. According to Gyr and Bewersdorff (1995) and Rosen (1989), surface tension behaviour of the solution depends upon the chemistry and concentration of the additive and temperature of the solution. Surfactants are essentially hydrophilic and hydrophobic chemical compounds with low molecular weight.

Tzan and Yang (1990) and Wang and Hartnett (1994) reported the effect of surfactant type and its concentration on surface tension of solution. Wu et al. (1995) reported the variation of surface tension of the aqueous solution with the concentration of surfactant solutions as shown in Fig. 5.25, and they found that surface tension decreased with increase in additive concentration. The addition of polymer concentration to the solvent increased the solution viscosity. It was found that for a given concentration, viscosity can also be increased with increase in the molecular weight of the polymer. Figure 5.26 depicts the variation of apparent viscosity with the shear rate for aqueous polymer solutions. They found that nucleate boiling heat transfer coefficient was increased with increase in the concentration of the surfactant in aqueous solution. Figure 5.27 shows the variation of boiling heat transfer coefficient with concentration of aqueous lauryl sulphate solution (SLS) at various heat fluxes. Cornwell and Houston, van Stralen and Cole (1979), Garg et al. (1980) and Chun and Kang (1998) experimented on pool boiling and reported that boiling behaviour was influenced by heated surface geometry and size.

Table 5.9 Literature study on aqueous solution (Wasekar and Manglik 2017)

Author(s)	Heater geometry and heat flux level (kW/m²)	Additives	Effect on heat transfer
Surfactant solutions:			
Morgan et al. (1949)	Cylinder 0–500	Drene (triethanolamine alkyl sulphate), and sodium lauryl sulfonate	Enhanced heat transfer with a maximum of around 100%
Jontz and Myers (1960)	Plate 20–25	Tergitol (sodium tetradecyl sulphate) and Aerosol-22 (n-octadecyl tetrasodium-1,2-dicarboxyethyl sulfosuccinamate)	Heat transfer coefficient enhancement to the extent of 50%
Podsushnyy et al. (1980)	Cylinder 0–80	PVS-6 polyvinyl alcohol, NP-3 sulfonol and SV 1017 wetting agent	Enhancement has an optimum value for concentration corresponding to cm. In higher heat flux range, the enhancement is higher for larger size heater
Filippov and Saltonov (1982)	Cylinder 0–100	Octadecylamine	Heat transfer coefficient improved by 100%.
Yang and Maa (1983)	Plate 0–600	SLS (sodium lauryl sulphate) and SLBS (sodium lauryl benzene sulfonate)	Maximum enhancement of 100–200% in the nucleate boiling regime, and a 50–100% increase in critical heat flux. Also, the enhancement depends on particular surfactant in solution
Saltanov et al. (1986)	Cylinder 40–120	Octadecylamine	Maximum enhancement to the extent of 100% for an optimum level of additive concentration
Tzan and Yang (1990)	Cylinder 0–400	SLS (sodium lauryl sulphate)	Enhanced with a maximum increase of around 200% with and optimum value of surfactant concentration at high heat fluxes (>300 kW/m²)
Liu et al. (1990)	Plate 0–400	BA-1, BA-2, BA-3, BA-4, DPE-1, DPE-3. Gelatine, oleic acid, trimethyl octadecyl ammonium chloride, trialkyl methyl ammonium chloride and polyvinyl alcohol	Maximum enhancement in the range of 200–700% observed with BA-1, BA-2 and BA-3, while no effect with other additives

(continued)

Table 5.9 (continued)

Author(s)	Heater geometry and heat flux level (kW/m²)	Additives	Effect on heat transfer
Chou and Yang (1991)	Plate 0–250	SLS (sodium lauryl sulphate)	Maximum enhancement of around 150%
Wu and Yang (1992)	Cylinder 23	SLS (sodium lauryl sulphate)	Decrease in incipient superheat and reduction in bubble size
Wang and Hartnett (1994)	Wire 0–600	SLS (sodium lauryl sulphate), and Tween-80 (polyoxyethylene sorbitan mono-oleate)	Heal transfer performance with SLS is similar to that of pure water; with Tween-80, the performance is slightly inferior
Wu et al. (1995)	Cylinder 0–450	SLS (sodium lauryl sulphate), Aerosol-2 2 (n-Octadecyl tetrasodium-1,2-dicarboxyethyl sulfosuccinamate), Tergitol (sodium tetradecylsulphate), DTMAC, Tween-20, Tween-40, Tween-80, n-octanol and triton X-100	Extent of enhancement depends on the surfactant with maximum enhancement (~100%) observed with SLS and Tergitol
Ammerman and You (1996)	Wire 0–450	SLS (sodium lauryl sulphate)	Enhancement up to a maximum of 50% with relative increase of convective component and corresponding decrease of the latent heat component of heat flux
Manglik (1998)	Cylinder 0–100	AGS (alkyl glyceryl sulfonate)	Enhancement to the extent of 100% with an optimum value of concentration for maximum heat transfer. Significantly early departure from nucleate boiling at higher concentrations
Polymeric solutions:			
Kotchaphakdee and Williams (1970)	Plate 0–900	Acrylamide, polyacrylamides (PA-10, PA-20) and hydroxyethyl cellulose (HEC-L, HEC-M, HEC-H)	Maximum enhancement of 250%; found to be limited by solution viscosity
Miaw (1978)	Plate 0–900	Hydroxyethyl cellulose (HEC-H), and polyacrylamides (PA-10, PA-30)	Maximum enhancement of 100% was observed with HEC-H, but no enhancement with PA solutions
Yang and Maa (1982)	Plate and wire 0–600	Hydroxyethyl cellulose and Natrosol (250HR, 300HR, 250GR)	No enhancement and no effect of heater geometry

Paul and Abdel Khalik (1983)	Wire 0–500	Polyacrylamides (AP-30, NP-IOP, MGL), hydroxyethyl cellulose (HHR, HR, MR), Galactomannan polysaccharide and polyethylene oxides	Decrease in heat transfer, with the solution viscosity itself correlating the corresponding decrease
Ulicny (1984)	Plate 0–220	Hydroxyethyl cellulose and polyacrylamide (PA-10)	Heat transfer enhanced by 200%, and the performance has a heater surface dependence
Hu (1989)	Wire 0–550	Polyacrylamide (Separan AP-30), hydroxyethyl cellulose (Natrosol 250HHR)	Enhancement observed with HEC at high heat fluxes and for higher concentrations (> 1000 wppm) while PA solutions show reduction in heat transfer
		Surfactant and polymer solutions:	
Wang and Hartnett (1992)	Wire 0–500	SLS (sodium lauryl sulphate) and polyacrylamide (AP-30)	Maximum enhancement of up to 100% observed when SLS is added to aqueous AP-30 solution at high heat fluxes (>200 kW/m^2)

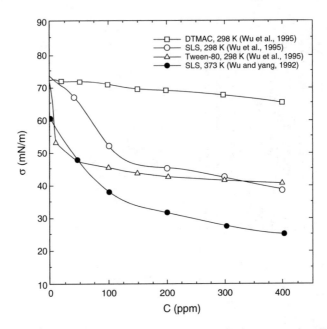

Fig. 5.25 Variation of surface tension of aqueous solution with concentration (Wasekar and Manglik 2017)

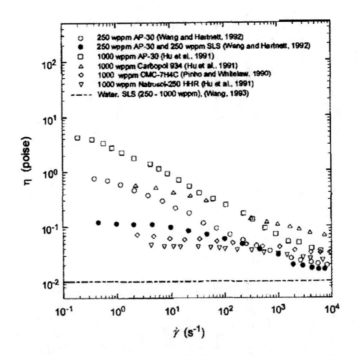

Fig. 5.26 Apparent shear viscosity variations with shear rates for an aqueous polymer solution (Wasekar and Manglik 2017)

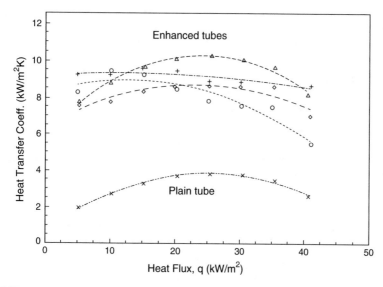

Fig. 5.27 Falling film heat transfer coefficients for R-134a on single enhanced tubes and plain tube [o-Turbo-B, Δ-Turbo-Cii, ◊-Wieland-SE, + Wieland-SC, ×-Plain] (Thome 2017)

References

Akbarinia A, Behzadmehr A (2007) Numerical study of laminar mixed convection of a Nanofluid in horizontal curved tubes. Appl Thermal Eng 27:1327–1337

Akbari OA, Toghraie D, Karimipour A, Safaei MR, Goodarzi M, Alipour H, Dahari M (2016) Investigation of rib's height effect on heat transfer and flow parameters of laminar water–Al_2O_3 nanofluid in a rib-microchannel. Appl Mathemat Comput 290:135–153

Ali HM, Generous MM, Ahmad F, Irfan M (2017) Experimental investigation of nucleate pool boiling heat transfer enhancement of TiO2-water based nano-fluids. Appl Therm Eng 113:1146–1151

Allen PHG, Cooper P (1987) The potential of electrically enhanced evaporators Third international symposium on the large scale application of heat pumps, Oxford, UK 221–229

Ammerman CN, You SM (1996) Determination of the boiling enhancement mechanism caused by surfactant addition to water. J Heat Transf 118:429–435

Arani AAA, Akbari OA, Safaei MR, Marzban A, Alrashed AA, Ahmadi GR, Nguyen TK (2017) Heat transfer improvement of water/single-wall carbon nanotubes (SWCNT) nanofluid in a novel design of a truncated double-layered microchannel heat sink. Int J Heat Mass Trans 113:780–795

Avila R, Cervantes J (1995) Analysis of the heat transfer coefficient in a turbulent particle pipe flow. Int J Heat Mass Transf 38(11):1923–1932

Babcsán N, Mészáros I, Hegman N (2003) Thermal and electrical conductivity measurements on aluminum foams. Mat Wiss u Werkstofftech 34:391–394

Bang IC, Chang SH (2004) Boiling heat transfer performance and phenomena of Al_2O_3– water Nanofluids from a plain surface in a pool. Int J Heat Mass Transf 48:2407–2419

Bartel WJ, Genetti WE (1973) Heat transfer from a horizontal bundle of bare and finned tubes in an air fluidized bed. A!ChE Svmp Ser 69(128):85–92

Bergles AE, Scarola LS (1966) Effect of a volatile additive on the critical heat flux for surface boiling of water in tubes. Chem Eng Sci 21:721–723

Bhatti MS, Savery SW (1975) Augmentation of heat transfer in a laminar external gas boundary layer by the vaporization of suspended droplets. J Heat Transf 97:179–184

Boothroyd RG, Haque H (1970) Fully developed heat transfer to a gaseous suspension of particles flowing turbulently in duct of different size. J Mech Eng Sci 12(3):191–200

Bonilla CF, Cervi A Jr, Colven TJ Jr, Wang SJ (1953) Heat transfer to slurries in pipe, chalk, and water in turbulent flow. A!ChE Symp Sen 49(5):127–134

Cheedarala RK, Park E, Kong K, Park YB, Park HW (2016) Experimental study on critical heat flux of highly efficient soft hydrophilic CuO–chitosan nano-fluid templates. Int J Heat Mass Transf 100:396–406

Chein R, Chuang J (2007) Experimental microchannel heat sink performance studies using nanofluids. Int J Therm Sci 46(1):57–66

Cheol P, Zoubeida O, Watson KA, Crooks RE, Smith J, Lowther SE, Connell JW, Siochi EJ, Harrison JS, Clair TL (2002) Dispersion of single wall carbon nanotubes by in situ polymerization under sonication. Chem Phys Lett 364:303–308

Chen JC, Withers JG (1978) An experimental study of heat transfer from plain and finned tubes in fluidized beds. A!ChE Symp Ser 74(174):327–333

Chitra SR, Sendhilnathan S, Suresh S (2015) Investigation of heat transfer characteristics of Mgmnni/Diw-based Nanofluids for quenching in industrial applications. J Enhanc Heat Transf 22(1):1

Cho YI, Hartnett JP (1982) Non-Newtonian fluids in circular pipe flow. In: Advances in heat transfer, vol 15. Academic, New York, pp 59–141

Cho HJ, Kang IS, Kweon YC, Kim MH (1996) Study of the behavior of a bubble attached to a wall in a uniform electric field. Int J Multiphase Flow 22:909–922

Choi SUS (1995) Enhancing thermal conductivity of fluids with nanoparticles. In: Singer DA, Wang HP (eds) Developments and applications of non-Newtonian flows. ASME, New York, pp 99–105

Choi YJ, Kam DH, Jeong YH (2017) Analysis of CHF enhancement by magnetite nanoparticle deposition in the subcooled flow boiling region. Int J Heat Mass Transf 109:1191–1199

Chou CC, Yang YM (1991) Surfactant effects on the temperature profile within the superheated boundary layer and the mechanism of nucleate pool boiling. J Chinese Institute Chem Eng 22 (2):71–80

Chun MH, Kang MG (1998) Effects of heat exchanger tube parameters on nucleate pool boiling heat transfer. J Heat Transf 120:468–476

Ciloglu D (2017) An experimental investigation of nucleate pool boiling heat transfer of nanofluids from a hemispherical surface. Heat Transf Eng 38(10):919–930

Das S, Putra N, Thiesen P, Roetzel W (2003a) Temperature dependence of thermal conductivity enhancement for Nanofluids. J Heat Transf 125:567–574

Das SK, Putra N, Roetzel W (2003b) Pool boiling characteristics of Nano-fluids. Int J Heat Mass Transf 46:851–862

Das SK, Putra N, Roetzel W (2003c) Pool boiling of Nano-fluids on horizontal narrow tubes. Int J Multiphase Flow 29:1237–1247

Depew CA, Reisbig RL (1964) Vapor condensation on a horizontal tube using Teflon to promote dropwise condensation. Ind Eng Chem Process Design Dev. 11: 365-369.

Ding Y, Wen D (2005) Particle migration in a flow of nanoparticle suspensions. Powder Technol 149:84–92

Ding Y, Alias H, Wen D, Williams RA (2006) Heat transfer of aqueous suspensions of carbon nanotubes (CNT Nanofluids). Int J Heat Mass Transf 49:240–250

Dizaji S (2014) Heat transfer enhancement due to air bubble injection into a horizontal double pipe heat exchanger. Int J Automotive Eng 4(4):902–910

Eastman JA, Choi SUS, Li S, Thompson LJ (1997) Enhanced thermal conductivity through the development of Nanofluids. Proc Symp on Nanophase and Nanocomposite materials II, materials research society. 457:3–11

Eastman JA, Choi SUS, Li S, Yu W, Thompson LJ (2001) Anomalously increased effective thermal conductivities of ethylene glycol-based Nanofluids containing copper nanoparticles. Appl Phys Lett 78:718–720

Elimelech M, Gregory J, Jia X, Williams RA (1995) Particle deposition and aggregation: measurement, modeling and simulation. Butterworths, Oxford

Esmaeili M, Sadeghy K, Moghaddami M (2010) Heat transfer enhancement of wavy channels using Al_2O_3 nanoparticles. J Enhanc Heat Transf 17(2):139–151

Filippov GA, Saltanov GA (1982) Steam-liquid media heat-mass transfer and hydrodynamics with surface-active substance additives. Heat Transfer, Vol. 4. Hemisphere Publishing Corporation:443–447

Fossa M, Tagliafico LA (1995) Experimental heat transfer of drag-reducing polymer solutions in enhanced surface heat exchangers. Exp Thermal Fluid Sci 10:221–228

Funfschilling D, Li HZ (2006) Effects of the injection period on the rise velocity and shape of a bubble in a non-Newtonian fluid. Chem Eng Res Des 84(10):875–883

Furchi JCL, Goldstein L, Lombardi G, Mohseni M (1988) Heat transfer coefficients in flowing gas-solid suspensions, A!ChE Symp. Ser., 84(263), 26–30

Gabillet C, Colin C, Fabre J (2002) Experimental study of bubble injection in a turbulent boundary layer. Int J Multiphase Flow 28(4):553–578

Gannett HJ Jr, Williams MC (1971) Pool boiling in dilute nonaqueous polymer solutions. Int J Heat Mass Transfer 11:1001–1005

Garg NS, Shankar U, Tripathi G (1980) Pool boiling heat transfer from rotating horizontal cylinders. Indian J Technol 18:53–56

Grassi W, Testi D (2006) Heat transfer augmentation by ion injection in an annular duct. J Heat Transf Trans ASME 128:283–289

Gravndyan Q, Akbari OA, Toghraie D, Marzban A, Mashayekhi R, Karimi R, Pourfattah F (2017) The effect of aspect ratios of rib on the heat transfer and laminar water/TiO_2 nanofluid flow in a two-dimensional rectangular microchannel. J Mol Liq 236:254–265

Grenwal NS, Saxena SC (1979) Effect of surface ronghness on heat transfer from horizontal immersed tubes in a fluidized bed. J Heat Transf 101:397–403

Griffith P (1985) Condensation. Part 2: Dropwise condensation. In Handbook of heat transfer applications. McGraw-Hill, New York, Chap. 11.

Gyr A, Bewersdorff HW (1995) Drag reduction of turbulent flows by additives. Kluwer, Netherlands

Ham J, Kim H, Shin Y, Cho H (2017) Experimental investigation of pool boiling characteristics in Al_2O_3 nanofluid according to surface roughness and concentration. Int J Therm Sci 114:86–97

Hamilton RL, Crosser OK (1962) Thermal conductivity of heterogeneous two component systems. Ind Eng Chem Fundam 1:187–191

He Y, Li H, Hu Y, Wang X, Zhu J (2016) Boiling heat transfer characteristics of ethylene glycol and water mixture based ZnO nano-fluids in a cylindrical vessel. Int J Heat Mass Transf 98:611–615

Heidary H, Kermani MJ (2012) Heat transfer enhancement in a channel with block (s) effect and utilizing nano-fluid. Int J Therm Sci 57:163–171

Heris SZ, Esfahany MN, Etemad G (2006) Investigation of CuO/water nanofluid laminar convective heat transfer through a circular tube. J Enhanc Heat Transf 13(4):279–289

Hong K, Hong TK, Yang HS (2006) Thermal conductivity of Fe Nanofluids depending on the cluster size of nanoparticles. Appl Phys Lett 88:31901-1–31901-3

Hong T-K, Yang H-S (2005) Nanoparticle-dispersion-dependent thermal conductivity in nanofluids. J Korean Phys Soc 47:321

Hu RYZ (1989) Nucleate pool boiling from a horizontal wire in viscoelastic fluid. Ph.D. Thesis, University of Illinois at Chicago, Chicago

Ide H, Kimura R, Kawaji M (2007) Optical measurement of void fraction and bubble size distributions in a microchannel. Heat Transf Eng 28(8–9):713–719

Irvine TF Jr, Kami J (1987) Non-Newtonian fluid flow and heat transfer. In: Kakat S (ed) Handbook of single-phase convective heat transfer. Wiley-Interscience, New York, p 20

Jontz PD, Myers JE (1960) The effect of dynamic surface tension on nucleate boiling coefficients. AIChE J 6:34–38

Kamel MS, Lezsovits F, Hussein AM, Mahian O, Wongwises S (2018) Latest developments in boiling critical heat flux using nanofluids: a concise review. Int Commun Heat Mass Transf 98:59–66

Katto Y, Kawamura S (1981) Critical heat flux during natural convective boiling on uniformly heated tubes submerged in saturated liquid. JSME B 47(423):2186–2190

Kenning DBR, Kao YS (1972) Convective heat transfer to water containing bubbles: enhancement not dependent on thermocapillarity. Int J Heat Mass Transf 15:1709–1718

Khanafer K, Vafai K, Lightstone M (2003) Buoyancy-driven heat transfer enhancement in a two-dimensional enclosure utilizing Nanofluids. Int J Heat Mass Transf 46(19):3639–3653

Kim KJ, Kulankara S, Herold K, Miller C (1996) Heat transfer additives for use in high temperature applications. Proc Int Absorp Heat Pump Conf. Montreal, Canada, 1, 89-97.

Kim KJ, Lefsaker AM, Razani A, Stone A (2001) The effective use of heat transfer additives for steam condensation. Appl Thermal Eng 21:1863–1874

Kitagawa A, Kosuge K, Uchida K, Hagiwara Y (2008) Heat transfer enhancement for laminar natural convection along a vertical plate due to sub-millimeter-bubble injection. Exp Fluids 45 (3):473–484

Kitagawa A, Kitada K, Hagiwara Y (2010) Experimental study on turbulent natural convection heat transfer in water with sub-millimeter-bubble injection. Exp Fluids 49(3):613–622

Kofanov VI (1964) Heat transfer and hydraulic resistance in flowing liquid suspensions in piping. Int Chem Eng 4(3):426–430

Koo J, Kleinstreuer C (2005) Laminar Nanofluid flow in microheat-sinks. Int J Heat Mass Transf 48:2652–2661

Kosky PG (1976) Heat transfer to saturated mist flowing normally to a heated cylinder. Int J Heat Mass Transf 19:539–543

Kotchaphakdee P, Williams MC (1970) Enhancement of nucleate pool boiling with polymeric additives. Int J Heat Mass Transf 13:835–848

Kowsary F, Heyhat MM (2011) Numerical investigation into the heat transfer enhancement of Nanofluids using a nonhomogeneous model. J Enhanc Heat Transf 18(1):81–90

Krause WB, Peters AR (1983) Heat transfer from horizontal serrated finned tubes in an air-fluidized bed of uniformly sized particles. J Heat Transf 105:319–324

Kumada M, Chu R, Sato K (2002) Heat transfer enhancement and flow characteristics of drag-reducing surfactant aqueous solutions using the turbulent promoter. Proc 12th Int Heat Transfer Conf 4:129–134

Kurosaki Y, Murasaki T (1986) Study on heat transfer mechanism of a gas–solid suspension impinging jet (effect of particle size and thermal properties). Proc 8th Int Heat Transfer Conf 5:2587–2592

Kwak SD, Oh Y (2000) A study of bubble behavior and boiling heat transfer enhancement under electric field. Heat Transf Eng 21(4):33–45

Kweon YC, Kim MH, Cho HJ, Kang IS (1998) Study on the deformation and departure of a bubble attached to a wall in DC/AC electric fields. Int J Multiphase Flow 24:145–162

Lee WK, Vaseleski RC, Metzner AB (1974) Turbulent drag reduction in polymeric solutions containing suspended fibers. AIChE J 20:128–133

Li P, Kawaguchi Y, Daisaka H, Yabe A, Hishida K, Maeda M (2001) Heat transfer enhancement to the drag-reducing flow of surfactant solution in two-dimensional channel with mesh-screen inserts at the inlet. J Heat Transf 123:779–789

Li Q, Xuan Y (2000) Experimental investigation of transport properties of nanofluids. In: Buxuan W (ed) Heat transfer science & technology. Higher Education Press, Beijing, pp 757–762

Li X, Zhu D, Wang X, Wang N, Gao J (2009) Thermal conductivity enhancement for aqueous alumina nano-suspensions in the presence of surfactant. J Enhanc Heat Transf 16(2):93–102

Liao L, Liu Z, Bao R (2010) Forced convective flow drag and heat transfer characteristics of CuO nanoparticle suspensions and nanofluids in a small tube. J Enhanc Heat Transf 17(1):45–57

Liu T, Cai Z, Lin J (1990) Enhancement of nucleate boiling heat transfer with additives. In: Deng S-J (ed) Heat transfer enhancement and energy conservation. Hemisphere Publishing Corp, Washington, DC, pp 417–424

Liu ZH (2001) Enhancement of boiling heat transfer in restricted spaces in compact horizontal tube bundles. Heat Transf–Asian Res 30:394–401

Liu Y, Li R, Wang F, Yu H (2004) The effect of electrode polarity on EHD enhancement of boiling heat transfer in a vertical tube. Exp Therm Fluid Sci 29:601–608

Lv LC, Liu Z (2008) Boiling heat transfer characteristics in small vertical tubes submerged in saturated nanoparticle suspensions. J Enhanc Heat Transf 15(2):101–112

Lv LC, Liu Z (2010) Effects of nanoparticle parameters on thermal performance of the evaporator in a small capillary pumped loop using nanofluid. J Enhanc Heat Transf 17(4):343–352

Maïga SEB, Nguyen CT, Galanis N, Roy G (2004) Heat transfer enhancement in forced convection laminar tube flow by using nanofluids. In Proc Int Symp Adv Comput Heat Transf CHT04, April 19-24; Norway

Masuda H, Ebata A, Teramae K, Hishinuma N (1993) Alteration of thermal conductivity and viscosity of liquid by dispersing ultra-fine particles (dispersion of γ-Al2O3, SiO2 and TiO2 ultra-fine particles). Netsu Bussei (Japan) 7(4):227–233

Manglik RM (1998) Pool boiling characteristics of high concentration aqueous surfactant emulsions. Heat Trans 2:449–453

Miaw CB (1978) A study of heat transfer to dilute polymer solutions in nucleate pool boiling. Ph.D. Thesis University of Michigan, Ann Arbor

Miller AP, Moulton RW (1956) Heat transfer to liquid-solid suspensions in turbulent flow in pipes. Trend Eng:15–21

Monde M, Yamaji K (1990) Critical heat flux during natural circulation boiling in a vertical uniformly heated tube submerged in saturated liquid. Int J Heat Transf 2:111–116

Morgan AI, Bromley LA, Wilkie CR (1949) Effect of surface tension on heat transfer in boiling. Ind Eng Chem 41:2767–2769

Moyls AL, Sabersky RH (1975) Heat transfer to dilute asbestos dispersions in smooth and rough tubes. Lett Heat Mass Trans 2:293–302

Murray DB (1994) Local enhancement of heat transfer in a particulate cross flow—I. Heat transfer mechanisms. Int J Multiphase Flow 20(3):493–504

Murshed SMS, Leong KC, Yang C (2005) Enhanced thermal conductivity of TiO_2—water based nanofluids. Int J Ther Sci 44(4):367–373

Neve RS, Yan YY (1996) Enhancement of heat exchanger performance using combined electrohydrodynamic and passive methods. Int J Heat Fluid Flow 17:403–409

Nishikawa N, Takase H (1979) Effects of particle size and temperature difference on mist flow over a heated circular cylinder. J Heat Transf 101:705–711

Nouri NM, Sarreshtehdari A (2009) An experimental study on the effect of air bubble injection on the flow induced rotational hub. Exp Thermal Fluid Sci 33(2):386–392

Ogata J, Yabe A (1991) Augmentation of nucleate boiling heat transfer by applying electric fields: EHD behavior of boiling bubble. Proc ASME/JSME Therm Eng 3:41–46

Ogata J, Yabe A (1993) Augmentation of boiling heat transfer by utilizing the EHD effect -EHD behaviour of boiling bubbles and heat transfer characteristics. Int J Heat Mass Transf 36:783–791

Ökten K, Biyikoglu A (2018a) Effect of air bubble injection on the overall heat transfer coefficient. J Enhanc Heat Transf 25(3):195

Ökten K, Biyikoglu A (2018b) Effect of air bubble injection on the overall heat transfer coefficient. J Enhan Heat Transf 25(3)

Orr C, Dallavalle JM (1954) Heat transfer properties of liquid-solid suspensions. Chem Eng Prag Symp Ser 50(9):29–45

Pak BC, Cho YI (1998) Hydrodynamic and heat transfer study of dispersed fluids with submicron metallic oxide particles. ExpHeat Transf 2:151–170

Pal SK, Bhattacharyya S (2018) Enhanced heat transfer of Cu-water nanofluid in a channel with wall mounted blunt ribs. J Enhan Heat Trans 25(1)

Paper RA, Ohadi MM, Kumar A, Ansari AI (1993) Effect of electrode geometry on EHD-enhanced boiling of R-123/oil mixture. ASHRAE Trans 99:1237–1243

Paul DD, Abdel-Khalik SI (1983) Nucleate boiling in drag reducing polymer solutions. J Rheol 27 (1):59–76

Petrie JC, Freeby JA, Buckham JA (1968) In-bed heat exchangers. Chem Eng Prog 45–51

Prasher R, Bhattacharya P, Phelan PE (2006) Brownian-motion-based convective conductive model for the thermal conductivity of nanofluids. Trans ASME J Heat Transf 128:588–595

Podsushnyy AM, Minyev AN, Statsenko VN, Yakubovskiy YV (1980) Effect of surfactants and of scale formation on boiling heat transfer to sea water. Heat Trans–Soviet Res 12(2):113–114

Qi Y, Kawaguchi Y, Lin Z, Ewing M, Christensen RN, Zakin JL (2001) Enhanced heat transfer of drag reducing surfactant solutions with fluted tube-in-tube heat exchanger. Int J Heat Mass Transf 44:1495–1505

Raisee M, Moghaddami M (2008) Numerical investigation of laminar forced convection of nanofluids through circular pipes. J Enhanc Heat Transf 15(4):335–350

Rodríguéz-Perez MA, Reglero JA, Lehmhus D, Wichmann M, De Saja JA, Fernández A (2003) The transient plane source technique (TPS) to measure thermal conductivity and its potential as a tool to detect in-homogeneities in metal foams, Proc Int Conf advanced metallic materials, Smolenice, Slovakia, 5–7 November: 253–257

Rohsenow WM (1952) A method of correlating heat transfer data for surface boiling liquids. Trans ASME 74:969–979

Rosen MJ (1989) Surfactants and interfacial phenomena, 2nd edn. Wiley, New York

Roy G, Nguyen CT, Lajoie PR (2004) Numerical investigation of laminar flow and heat transfer in a radial flow cooling system with the use of nanofluids. Superlattices Microstructures 35 (3):497–511

Roy GC, Nguyen CT, Comeau M (2006) Numerical investigation of electronic component cooling enhancement using nanofluids in a radial flow cooling system. J Enhanc Heat Transf 13 (2):101–115

Rush WF (1968) Field testing of additives. In Symposium on absorption air conditioning, American Gas Association, Chicago, IL

Rush W, Wurum J, Perez-Blanco H (1991) A brief review of additives for absorption enhancement, vol 91. Absorp Heat Pump Conf, Tokyo, Japan, pp 183–187

Sadek SE (1972) Heat transfer to air-solids suspensions in turbulent flow. Ind Eng Chem Process Design Dev 11:133–135

Saffari H, Moosavi R, Gholami E, Nouri NM (2013) The effect of bubble on pressure drop reduction in helical coil. Exp Thermal Fluid Sci 51:251–256

Saltanov GA, Kukushkin AN, Solodov AP, Sotskov SA, Jakusheva EV, Chempik E (1986) Surfactant influence on heat transfer at boiling and condensation. Heat Trans. 1986, Hemisphere Publishing Corporation, Washington, DC, Vol. 5, pp. 2245–2250

Samaroo R, Kaur N, Itoh K, Lee T, Banerjee S, Kawaji M (2014) Turbulent flow characteristics in an annulus under air bubble injection and subcooled flow boiling conditions. Nucl Eng Des 268:203–214

Sandhu H, Gangacharyulu D, Singh MK (2018) Experimental investigations on the cooling performance of micro-channels using alumina nano-fluids with different base fluids. J Enhanc Heat Transf 25(3):283

Sarafraz MM, Kiani T, Hormozi F (2016) Critical heat flux and pool boiling heat transfer analysis of synthesized zirconia aqueous nano-fluids. Int Commun Heat Mass Transf 70:75–83

Sato Y, Deutsch E, Simonin O (1998) Direct numerical simulation of heat transfer by solid particles suspended in homogenous isotropic turbulence. Int J Heat Fluid Flow 19:187–192

Sulaiman MZ, Matsuo D, Enoki K, Okawa T (2016) Systematic measurements of heat transfer characteristics in saturated pool boiling of water-based nano-fluids. Int J Heat Mass Transf 102:264–276

Tamari M, Nishikawa K (1976) The stirring effect of bubbles upon the heat transfer to liquids. Heat Trans Japanese Res 5(2):31–44

Tanasawa I (1978) Dropwise condensation: the way to practical applications. Proc 6th Int Heat Transfer Conf 6:393–405

Thomas WC, Sunderland JE (1970) Heat transfer between a plane surface and air containing water droplets. Ind Eng Chem Fundam 9:368–374

Thome JR (2017) A review on falling film evaporation. J Enhanc Heat Transf 24:1–6

Tsai CY, Chien HT, Ding PP, Chan B, Luh TY, Chen PH (2004) Effect of structural character of gold nanoparticles in nanofluid on heat pipe thermal performance. Mater Lett 58(9):1461–1465

Tu JP, Dinh N, Theofanous T (2004) An experimental study of nanofluid boiling heat transfer. Proc. 6th Int. Symp. on Heat Transfer, Beijing China

Tzan YL, Yang YM (1990) Experimental study of surfactant effects on pool boiling heat transfer. J Heat Trans 112:207–212

Ulicny JC (1984) Nucleate pool boiling in dilute polymer solutions. Ph.D. Thesis, University of Michigan, Ann Arbor

van Stralen SJD (1959) Heat transfer to boiling binary liquid mixtures. B1: Chem Eng 4(Patt I):8–17; 4 (Part II), 78–82

van Stralen SJD, Cole R (1979) Boiling phenomena: physicochemical and engineering fundamentals. Hemisphere, Washington 1:49–50

van Wijk, WR, Vos AS, van Stralen SJD (1956) Heat transfer to boiling binary liquid mixtures. Chem. Eng. Sci., 5:68–80

Wang C-C, Chen C-K (2002) Combined free and forced convection film condensation on a finite-size horizontal wavy plate. J Heat Trans 124:573–576

Wang TAA, Hartnett JP (1992) Influence of surfactants on pool boiling of aqueous polyacrylamide solutions. Warme Stoffubertrag 27:245–248

Wang TA A, Hartnett JP (1994) Pool boiling heat transfer from a horizontal wire to aqueous surfactant solutions. Heat Transfer 1994, I Chem. E, UK, 5:177–182

Wasekar VM, Manglik RM (2017) Enhanced heat transfer in nucleate pool boiling of aqueous surfactant and polymeric solutions. J Enhanc Heat Transf 24(1-6)

Wasekar VM, Manglik RM (1999) A review of enhanced heat transfer in nucleate pool boiling of aqueous surfactant and polymeric solutions. J Enhan Heat Transf 6:135–150

Wasekar VM, Manglik RM (2000) Pool boiling heat transfer in aqueous solutions of an anionic surfactant. J Heat Transf 122:708–715

Watkins RW, Robertson CR, Acrivos A (1976) Entrance region heat transfer in flowing suspensions. Int J Heat Mass Transf 19:693–695

Webb RL, Kim NY (2005) Principles of enhanced heat transfer. Taylor and Francis, New York

Wen DS, Wang BX (2002) Effects of surface wettability on nucleate pool boiling heat transfer for surfactant solutions. Int J Heat Mass Transf 45:1739–1747

Wen D, Ding Y (2004) Experimental investigation into convective heat transfer of Nanofluids at the entrance region under laminar flow conditions. Int J Heat Mass Transf 47:5181–5188

Wen D, Ding Y, Williams RA (2006) Pool boiling heat transfer of aqueous TiO_2-based nanofluids. J Enhanc Heat Transf 13(3):231–244

Witharana S (2003) Boiling of refrigerants on enhanced surfaces and boiling of Nanofluids. PhD Thesis, Royal Institute of Technology, Stockholm, Sweden

Wu WT, Yang YM, Maa JR (1995) Enhancement of nucleate boiling heat transfer and depression of surface tension by surfactant additives. J Heat Transf 117:526–529

Wu W-T, Yang Y-M (1992) Enhanced boiling heat transfer by surfactant additives. In Dhir VK, Bergles AE (eds) Proceedings of the engineering foundation conference on pool and external flow boiling. Santa Barbara, CA, 361–366

Wu W-T, Yang Y-M, Maa J-R (1998) Nucleate pool boiling enhancement by means of surfactant additives. Exp Thermal Fluid Sci 18:195–209

Xuan Y, Li Q (2000) Heat transfer enhancement of Nanofluids. Int J Heat Fluid Flow 21(1):58–64

Xuan Y, Li Q (2003) Investigation on convective heat transfer and flow features of nanofluids. J Heat Transf 125:151–155

Xuan Y, Roetzel W (2000) Conception for heat transfer correlation of nanofluid. Int J Heat Mass Transf 43(19):3701–3707

Yang YM, Maa JR (1982) Effects of polymer additives on pool boiling phenomena. Letters in Heat Mass Transfer 9:237–244

Yang YM, Maa JR (1983) Pool boiling of dilute surfactant solutions. J Heat Trans 105:190–192

Yang Y-M, Lin C-Y, Liu M-H, Maa J-R (2002) Lower limit of the possible nucleate pool boiling enhancement by surfactant addition to water. J Enhan Heat Transf 9:153–160

Yoo SS (1974) Heat transfer and friction factor for nonnewtonian fluids in turbulent pipe flow. Ph.D Thesis, University of Illinois, Chicago

You SM, Kim JH, Kim KH (2003) Effect of nanoparticles on critical heat flux of water in Pool boiling heat transfer. Appl Phys Lett 83:3374–3376

Yu W, France DM, Routbort JL, Choi SUS (2008) Review and comparison of nanofluid thermal conductivity and heat transfer enhancements. Heat Transf Eng 29(5):432–460

Zeinali Heris S, Nasr Esfahany M, Etemad SG (2005) Experimental investigation of convective heat transfer of Nanofluid in circular tube. Int J Heat Fluid Flow 28(2):203–210

Zeinali Heris S, Nasr Esfahany M, Etemad SG (2007) Experimental investigation of convective heat transfer of Al_2O_3/water Nanofluid in circular tube. Int J Heat Fluid Flow 28:203–210

Zhang S, Luo Z, Wang T, Shou C, Ni M, Cen K (2010) Experimental study on the convective heat transfer of CuO– water Nanofluid in a turbulent flow. J Enhanc Heat Transf 17(2):183–196

Zhu DS, Li XF, Wang XJ (2007) Study on preparation and dispersion behavior of Al_2O_3–H_2O nanofluids. Chinese J New Chem Mater 35:45–47

Ziegler F, Grossman G (1996) Heat transfer enhancement by additives. Int J Refrig 19:301–309

Chapter 6
Conclusions

Following conclusions are drawn.

- EHD enhancement of boiling and condensation has been discussed in detail. This enhancement technique is useful for practical purposes.
- More research is needed to compare the EHD performance vis-à-vis other enhancement techniques.
- Heat transfer enhancement in simultaneous heat–mass transfer process has been discussed. The enhancement of gas phase or liquid phase or both the phases has to be considered. Mass transfer during humidification and dehumidification is a subject of involved study in which heat–mass transfer analogy is often drawn.
- Cooling towers use enhanced surface geometries.
- Water film enhancement in fin-and-tube heat exchangers used for heat rejection is of interest, and surface wetting is a matter of serious investigation.
- Mass transfer in both the liquid and the gas phases need to be investigated more in detail.
- Typical solid–liquid suspension is not of much use. Solid–gas suspensions, fluidized beds with finned surfaces, are of interest.
- Aqueous dilute polymer additives are useful.
- Minute amount of a volatile fluid and addition of a surfactant enhance nucleate boiling heat transfer of water.
- Additives promote high-performance dropwise condensation depending upon surface tension of the fluid.

© The Author(s), under exclusive license to Springer Nature Switzerland AG 2020

S. K. Saha et al., *Electric Fields, Additives and Simultaneous Heat and Mass Transfer in Heat Transfer Enhancement*, SpringerBriefs in Applied Sciences and Technology, https://doi.org/10.1007/978-3-030-20773-1_6

Additional References

Bergles AE, Junkhan GH, Hagge JK (1976) Advanced cooling Systems for Agricultural and Industrial Machines. SAE paper 7 51183, Warrendale, PA

Boreyko JB, Chen CH (2009) Self-propelled dropwise condensate on superhydrophobic surfaces. Phys Rev Lett 103(18):184501

Chen F, Liu D, Song Y, Peng Y (2011) Visualization of a single boiling bubble in a DC electric field. In: ASME-JSME-KSME 2011 joint fluids engineering conference. American Society of Mechanical Engineers, pp 2733–2739

Cheung K, Ohadi MM, Singh A (1997) EHD-enhanced boiling coefficients and visualization of R-134a over enhanced tubes. J Heat Transfer 119:332–338

Choi HY, Reynolds JM (1965) Study of electrostatic fields on condensing heat transfer, air force dynamics laboratory report AFFDL-TR-65-51. Wright-Patterson Air Force Base, Ohio

Comini G, Croce G (2001) Convective heat and mass transfer in tube-fin exchangers under dehumidifying conditions. Numerical Heat Transf Part A Appl 40:579–599

Da Silva L, Molki M, Ohadi, MM (2001) Effect of polarity on electrohydrodynamic enhancement of R-134a condensation on enhanced tubes., in proceedings of 35th National Heat Transfer Conference, Anaheim, CA, NHTC 2001–20067

Depew CA, Kramer TJ (1973) Heat transfer to flowing gas-solid mixtures. In: Hartnett JP, Irvine TF (eds) Advances in heat transfer, vol 9. Academic Press, New York, pp 113–180

Herold K, Raderrnacher R, Klein SA (1996) Absorption chillers and heat pumps. CRC Press, Boca Raton, FL

Hetsroni G, Gurevich M, Mosyak A, Rozenblit R, Yarin LP (2002) Subcooled boiling of surfactant solutions. Int J Heat Mass Transf 28:347–361

Hetsroni G, Gurevich M, Mosyak A, Pogrebnyak E, Rozenblit R, Segal Z (2004) The effect of surfactants on boiling heat transfer. J Enhanc Heat Transf 13(2):185–195

Li Q, Xuan Y (2002) Convective heat transfer performance of fluids with nano-particles. In: Proc 12th Int. heat transfer Conf, vol 1, pp 483–488

Micic BB, Rohsenow WM (1969) A new correlation of pool boiling data including the effect of heating surface characteristics. J Heat Transfer 83:245–250

Molki M, Ohadi MM, Bloshteyn M (2000a) Frost reduction under intermittent electric field, proceedings of 34" National Heat Transfer Conference, Pittsburgh, PA, NHTC 2000–12052

© The Author(s), under exclusive license to Springer Nature Switzerland AG 2020
S. K. Saha et al., *Electric Fields, Additives and Simultaneous Heat and Mass Transfer in Heat Transfer Enhancement*, SpringerBriefs in Applied Sciences and Technology, https://doi.org/10.1007/978-3-030-20773-1

Molki M, Ohadi MM, Baumgarten B, Hasegawa M, Yabe A (2000b) Heat transfer enhancement of airflow in a channel using corona discharge. J Enhanced Heat Transfer 7:411

Panofsky W, Phillips M (1962) Classical electricity and magnetism, 2nd edn. Addison-Wesley, Reading, MA, pp 107–116

Papar RA, Ohadi MM, Kumar A, Ansari AI (1993) Effect of electrode geometry on EHD enhanced boiling of R-123/oil mixture. ASHRAE Trans 99(Part 1):1237–1243

Sato K, Mimatsu J, Kumada M (1999) Drag reduction and heat transfer augmentation of swfactant additives in two dimensional channel flow. In: Proc. 5th ASME/JSME thermal Eng. Joint Conf. paper ATJE99, p 6452

Seshimo Y, Ogawa K, Marumoto K, Fujii M (1990) Heat and mass transfer performances on plate fin and tube heat exchangers under dehumidification. Heat Transf - Jpn Res (USA) 18(5):716–721

Sun XY, Dai YJ, Ge TS, Zhao Y, Wang RZ (2019) Heat and mass transfer comparisons of desiccant coated microchannel and fin-and-tube heat exchangers. Appl Therm Eng 150:1159

Thome JR (1990) Enhanced boiling heat transfer. Hemisphere, Washington, DC

Wang XQ, Mujumdar AS (2007) Heat transfer characteristics of nanofluids: a review. Int J Therm Sci 46(1):1–19

Webb RL, Eckert ERG, Goldstein RJ (1970) Heat transfer and friction in tubes with repeatedrib roughness. Int J Heat Mass Transf 14:601–617

Xia Y, Jacobi AM (2005) Air-side data interpretation and performance analysis for heat exchangers with simultaneous heat and mass transfer: wet and frosted surfaces. Int J Heat Mass Transf 48:5089–5102

Zaghdoudi MC, Lallemand M (2001) Nucleate pool boiling under DC electric field. Exp Heat Transfer 14(3):157–180

Zhang G, Wang B, Li X, Shi W, Cao Y (2018) Review of experimentation and modeling of heat and mass transfer performance of fin-and-tube heat exchangers with dehumidification. Appl Therm Eng

Index

© The Author(s), under exclusive license to Springer Nature Switzerland AG 2020
S. K. Saha et al., *Electric Fields, Additives and Simultaneous Heat and Mass
Transfer in Heat Transfer Enhancement*, SpringerBriefs in Applied Sciences
and Technology, https://doi.org/10.1007/978-3-030-20773-1

Printed in the United States
By Bookmasters